Python for Engineers and Scientists

This book focuses on the basics of Python programming fundamentals. It provides an introduction to present-day applications in technology and the upcoming state-of-art trends in a comprehensive manner. It is based on Python 3.x, and it covers the fundamentals of Python with object-oriented concepts having numerous worked out examples. This book also acts as a learning tool for students of beginner level as well as for researchers of advanced level. Each chapter contains numerous programming examples along with their outputs that explain the usage of the methods and functions discussed in the chapter.

This book

- Includes programming tips to highlight the important concepts and help readers avoid common programming errors.
- Provides programming examples along with their outputs to ensure their correctness and help readers to master the art of writing efficient Python programs.
- Contains MCQs (multiple choice questions) with their answers, conceptual questions and programming questions; and solutions to some selected programming questions, for every chapter.
- Discusses applications like time zone converter and password generators at the end.
- Covers the fundamentals of Python up to object-oriented concepts including regular expression.

This book offers a simple and lucid treatment of concepts supported with illustrations for easy understanding, provides numerous programming examples along with their outputs, and includes programming tips to highlight the important concepts. It is a valuable resource for senior undergraduates, graduate students and professionals in the fields of electrical engineering, electronics and communication engineering and computer engineering.

Python for Engineers and Scientists

Concepts and Applications

Rakesh Nayak and Nishu Gupta

CRC Press
Taylor & Francis Group
Boca Raton London New York

CRC Press is an imprint of the
Taylor & Francis Group, an **informa** business

First edition published 2023
by CRC Press
6000 Broken Sound Parkway NW, Suite 300, Boca Raton, FL 33487-2742

and by CRC Press
4 Park Square, Milton Park, Abingdon, Oxon, OX14 4RN

CRC Press is an imprint of Taylor & Francis Group, LLC

© 2023 Rakesh Nayak and Nishu Gupta

Library of Congress Cataloging-in-Publication Data

Names: Nayak, Rakesh, author. | Gupta, Nishu, author.
Title: Python for engineers and scientists : concepts and applications /
   Rakesh Nayak and Nishu Gupta.
Description: First edition. | Boca Raton : CRC Press, 2023. | Includes
   bibliographical references and index.
Identifiers: LCCN 2022028099 (print) | LCCN 2022028100 (ebook) | ISBN
   9781032111032 (hardback) | ISBN 9781032112596 (paperback) | ISBN
   9781003219125 (ebook)
Subjects: LCSH: Python (Computer program language) | Engineering--Data
   processing. | Science--Data processing. | Computer programming.
Classification: LCC QA76.73.P98 N93 2023 (print) | LCC QA76.73.P98
   (ebook) | DDC 005.13/3--dc23/eng/20220805
LC record available at https://lccn.loc.gov/2022028099
LC ebook record available at https://lccn.loc.gov/2022028100

ISBN: 9781032111032 (hbk)
ISBN: 9781032112596 (pbk)
ISBN: 9781003219125 (ebk)
ISBN: 9781032390970 (ebk plus)

DOI: 10.1201/9781003219125

Typeset in Times
by Deanta Global Publishing Services, Chennai, India

Prof. Rakesh Nayak dedicates this book to his late brother, Mr. Rajesh Nayak.

Dr. Nishu Gupta dedicates this book to his father, Professor K.M. Gupta.

Contents

SECTION I   Python Fundamentals

SECTION II   Object Oriented Concepts in Python

Foreword

I am pleased to write this foreword because I feel the book, *Python for Engineers and Scientists: Concepts and Applications*, deeply emphasizes the state-of-the-art technologies that comprise many research explorations in the field of artificial intelligence and that it can offer in taking care of various applications through networking and automation.

Although the focus of this work is on Python programming, it contains much more that will be of interest to those outside this subject. The range of topics covered in this book is quite extensive. The examples are very handy, with discussions that focus on step-wise learning, hands-on practice and concept building from the basic level, making learning very comfortable, yet professional. In this book, I see that the authors have conceptualized the intellectual foundations of artificial intelligence, elaborated its distinctive pedagogy in diversified technological applications and studied its patterns and impact on the common man. It will certainly help researchers and professional practitioners develop a shared vision and understanding of the interpretive discussion.

Python is a fun and extremely easy-to-use programming language that has steadily gained in popularity over the last few years. Python is well suited for rapid application development. Python is also an excellent cross-platform language. Because it's interpreted, the same code can run on any platform that has a Python interpreter, and almost all current platforms have one. The authors have made the contents of the chapters so easy to understand which helps beginners to gain confidence and gradually turns them into experts.

We encounter challenges in addressing technological problems. These challenges are both difficult and interesting. Researchers are working on them to develop new approaches and provide new solutions to keep up with the ever-changing potential threats. This book is a good initiative in that direction. It provides a solid platform for various realms of existing and upcoming technologies. The authors can be confident that there will be many grateful readers who will have gained a broader perspective of the disciplines of Python programming and its applications because of their efforts. I hope this book will serve as a primer for industry and academia, professional developers and upcoming researchers across the globe to learn, innovate and realize the multi-fold capabilities of scientific and engineering applications. Although Python is still a young and evolving language, I believe that it has a bright future in education. This book is an important step in that direction.

I wish good luck to the authors and readers of the book.

Dr. Ariel Soales Teres
Professor
Department of Research, Graduate and Innovation
Federal Institute of Maranhão, Brazil

Foreword

Foreword

I am delighted to write the foreword for this book, *Python for Engineers and Scientists: Concepts and Applications*. This book highlights the importance of the Python programming language, and what it offers in solving many real-world problems. It demonstrates to its readers useful concepts and applications that cater to improved technological solutions. The contents of the book are very promising and focus on the core functionality of the application by taking care of common programming tasks.

The contents of this book seem to have made Python programming a trending junction where both general-purpose programming and web applications can be accomplished. Also, one can use Python for developing complex scientific and numeric applications by referring to this book. It is so nicely designed with features to facilitate data analysis and visualization. One can take advantage of the data analysis features of Python to create custom big data solutions without putting in extra time and effort. At the same time, the data visualization libraries and APIs provided by Python help to visualize and present data in a more appealing and effective way. I am pretty sure that this book will help many researchers and learners extensively to solve artificial intelligence and natural language processing tasks.

The range of topics covered in this book is quite extensive and every topic is discussed by experts in their own field. Overall, the book provides a window to the research and development in the field of Python programming in a comprehensive way and enumerates the evolutions of contributing tools and techniques.

I highly recommend this book to a variety of audiences, including academicians, commercial engineers, researchers, students and scholars. It is my desire and expectation that this book will provide an effective learning experience, a contemporary update and a practical reference for all those who are interested in this versatile and trending programming language.

Prof. Banshidhar Majhi
Vice-Chancellor,
VSSUT Burla, Odisha, India
(On Lien from NIT Rourkela, Odisha, India)

Preface

Python is a simple yet powerful programming language with excellent functionality for processing data. Python is extensively used in industry, scientific research and technical education around the world. Python is often praised for the way it facilitates productivity, quality and maintainability of the software.

Written by experts, this book intends to present to its readers the concepts of the Python programming language. It combines fundamental theoretical discussions, examples, exercises, questions and activities for all core concepts. It covers a wide audience including university professors, graduate and Ph.D. students, industry professionals and researchers. Readers can learn to use Python best practices to solve real-world problems. The exercises and activities have been chosen specifically to grasp the concepts covered and enhance practical learning. The book is targeted not only at beginners but will also be equally beneficial to professional developers who are not yet familiar with Python. Experienced programmers can quickly learn enough Python using this book to get immersed in further exploration.

The chapters of the book have been organized into a reader-friendly manner and presented under two main sections: *Python Fundamentals*, suitable for readers with no prior knowledge of programming and *Object-Oriented Concepts in Python*, for readers who carry prior experience of OOPs programming languages. The first part is dedicated to the introduction and overview of programming fundamentals and the second part focuses on classes and objects, inheritance, polymorphism, aggregation and exception handling.

We believe this book is unique in providing a comprehensive framework for students to learn about Python in the context of learning to program. What sets these materials apart is the tight coupling of the chapters and exercises, giving even those with no prior programming experience a practical introduction to this language.

Prof. Rakesh Nayak

Dr. Nishu Gupta

Acknowledgments

I, Prof. Rakesh Nayak, wish to acknowledge the support and patience of my parents, Mrs. Jyotsnamayee and Mr. Manabhanjan Nayak and my wife, Mrs. Tusarika Nayak. I wish to thank my nephew, Mr. Sushobhan Raj Nayak, for his efforts while preparing the manuscript. I also wish to thank my late brother, Mr. Rajesh Nayak, who always encouraged me to achieve greater heights in life. I miss you Dada...

I am highly grateful to my Ph.D. supervisors Dr. C. V. Sastry, Computer Science Department, National Institute of Technology Rourkela (Formerly Regional Engineering College), Odisha, India, and Prof. Jayaram Pradhan, Computer Science Department, Behrampur University, Odisha, India, whose guidance has always encouraged me.

I, Dr. Nishu Gupta, acknowledge the inspiration and blessings of my mother Smt. Rita Rani Gupta and father Prof. K.M. Gupta. I am full of gratitude to my sister Mrs. Nidhi Gupta, brother-in-law CA Ritesh Shankar Gupta, wife Smt. Anamika Gupta, son Ayaansh Gupta and other family members for the patience shown and encouragement given to complete this venture. I deeply acknowledge the blessings of my academic advisor and mentor Prof. Rajeev Tripathi, Motilal Nehru National Institute of Technology Allahabad, Uttar Pradesh, India. I am highly grateful to my Ph.D. supervisor Prof. Arun Prakash, Electronics and Communication Engineering Department, MNNIT Allahabad, India, whose guidance has always encouraged me to do my best. I profoundly acknowledge the cooperation and encouragement showered on me by Prof. Rakesh Nayak, Department of Computer Science and Engineering, O.P. Jindal University, Raigarh, Chhattisgarh, India. He has been a major driving force in bringing out this book. I wholeheartedly acknowledge the motivation given to me by Dr. Anil Gupta, Centre for Development of Advanced Computing (C-DAC), Pune, Maharashtra, India; Dr. Krishan Kumar, National Institute of Technology Hamirpur, Himachal Pradesh, India; Prof. Mohammad Derawi, Norwegian University of Science and Technology (NTNU) Gjøvik, Norway; Mr. Gaurvendra Singh, IIFL Wealth, New Delhi, India; Er. Jalaj Kumar Singh, KPMG Global Services, India; Ms. Isha Bharti, Capgemini Inc., USA, and other colleagues and friends for their support and motivation in several ways.

We, the authors, wish to record our appreciation for the patience and endurance of our family members without whose support, this book would not have been possible.

We express our heartfelt gratitude to the publishers and their teams for their continued support and cooperation in publishing this book.

Authors

Dr. Rakesh Nayak, author of two text books, is currently a professor in the Department of Computer Science and Engineering at O. P. Jindal University, Raigarh, Chhattisgarh, India. He earned his Master's degree in Computer Applications from Indira Gandhi National Open University in the year 2007 and MTech in computer science and engineering from Acharya Nagarjuna University, Andhra Pradesh, India in 2010 and his PhD degree in Computer Science from Behrampur University, Odisha, India in 2013. Prior to joining the computer science department of O. P. Jindal University in January 2022, he worked in various capacities in different Engineering/MCA colleges. He has more than 22 years of teaching experience and has guided 11 MTech students. He has many publications in international journals to his credit.

Dr. Nishu Gupta is a senior member, IEEE. He is a postdoctoral fellow in the Smart Wireless Systems (SWS) Research Group at Norwegian University of Science and Technology (NTNU), located in Gjøvik, Norway. Before this position, he was an assistant professor in the Electronics and Communication Engineering department, College of Engineering and Technology, SRM Institute of Science and Technology, Chennai, India. He earned his PhD degree from Motilal Nehru National Institute of Technology Allahabad, Prayagraj, India, which is an Institute of National Importance as declared by the Govt. of India. He has authored and edited several books with international publishers such as Taylor & Francis, Springer, Wiley, Scrivener, among others. Dr. Nishu is on the editorial board of various International reputed journals and transactions. Dr. Nishu serves as an active reviewer in various highly reputed journals such as *IEEE Transactions on ITS, IEEE Access, IET Communications* and many more. He is a recipient of the Best Paper Presentation Award during an International Conference at Nanyang Technological University, Singapore. His research interests include Autonomous Vehicles, Edge Computing, Augmented Intelligence, Internet of Things, Internet of Vehicles, Deep Learning, Machine Learning, Ad-Hoc Networks, Vehicular Communication, Driving Efficiency, Cognitive Computing, and Human-Machine Interaction.

Section I

Python Fundamentals

1 Interaction with Python

After studying this chapter, the reader will be able to:

- Distinguish between identifiers and variables
- Understand various built-in datatypes and concepts of sequence
- Understand constants
- Write simple statements in a Python program
- Provide inputs to a Python program
- Get output from a Python program after execution
- Know different strings and number formatting statements
- Comment codes in a Python program

1.1 INTRODUCTION

We write a program to get at least one output when providing zero or more inputs. Interaction with a computer refers to two-way communication with it. These concepts may be entirely new to first time programmers. There are a few concepts which are common to all the programming languages while some may be specific to Python. These concepts are fundamental to understanding this book. The best way to learn is by writing the programs and executing them on our own.

1.2 VARIABLES AND IDENTIFIERS

A variable is a reserved location in a computer's memory to store values. The user can store different datatypes such as integers, floating point numbers and strings in that memory location. Python automatically reserves the memory space for each variable. The type of variable to be assigned depends on the data it contains. That is why Python is known as a *dynamic typing language*. This is in contrast to some programming languages such as *C*. Whenever a variable is to be used, it has to be defined and allocated space before it can be used whereas in Python, variables need not be declared before using them.

An identifier is a name given to a variable or an entity such as functions, objects, etc. Variables and identifiers are not the same. The name of the variable is an identifier but a variable has other properties such as value, type, scope, etc.

DOI: 10.1201/9781003219125-2

3

Example 1.1

Think of a person as a variable. A person has a name which is an identifier but that person also has other properties such as height, weight, eye color and other properties which are variables.

1.2.1   How to Name Identifiers

Just as names are assigned to persons, animals and plants to identify them, in a similar manner, in programming, names are given to variables, functions and objects to identify them.

Rules for Identifier:
1. The first thing that we have to remember is that identifiers are case sensitive. So identifier X and identifier x are not the same.
2. Identifiers can only contain letters, numbers and underscores, nothing else. For example, inside an identifier name, we cannot use a question mark, plus sign or multiplication sign. However, we can name an identifier *hello_there* and assign some value to it.
3. Identifiers cannot start with a number. That is, we can name an identifier as *x1*, but we cannot name it as *1x*. If we name an identifier *1x* and try to assign some value to it, Python gives an error called a *syntax error* which basically means it is an unacceptable syntax.
4. Identifier names cannot be the same as Python keywords. *Keywords* are the words defined in a Python environment to have a particular meaning. So we cannot use any of these words as our identifier. For example, we cannot define a variable named *while*.

Example 1.2

Identifier Name	Valid/Invalid	Reason
2abc	Invalid	Starts with numeric
_abc	Valid	An identifier can start with an underscore
ab'c	Invalid	Inverted comma symbol is not allowed
while	Invalid	It is a keyword

1.2.2   Use of Descriptive Identifiers and Comments

It is recommended to use descriptive names while naming variables. Moreover, if identifier names are selected such that they have more than one word, an underscore should be used to separate those words.

Example 1.3

```
Write a Python program to calculate the perimeter of a
rectangle. Assume length = 5 and breadth = 10.
x = 5
y = 10
z = 2 * (x + y)
print("The Perimeter of rectangle is: ", z)
```

We will get the perimeter to be 30. Look at the next program.

Example 1.4

```
Length = 5
Breadth = 10
Perimeter = 2 * ( Length + Breadth )
print("The Perimeter of rectangle is : ", Perimeter)
```

In this program, we get the same output as in the previous example. However, when we read this program, it is much easier to understand as the variable name is descriptive. When we have programs which are much larger and multiple variables are defined, it is always better to give each variable a more descriptive name. It leads to better debugging and scope of corrections in the future.

Example 1.5

```
Rectangle_Length = 5
Rectangle_Breadth = 10
Rectangle_Perimeter = 2 * (Rectangle_Length + Rectangle_Breadth)
print("The Perimeter of rectangle is : ", Rectangle_Perimeter)
```

In fact, while looking at the program, it should be easily understandable and the purpose of each line should be self-explanatory.

1.2.3 VARIABLE TYPES

We can convert everything to data. The datatype of a value is an attribute that tells what kind of data the variable value can have. The variety of available datatypes allows the programmer to select the appropriate one according to the needs of the application. In Python, each datatype is a *class*. We will discuss more about *classes* in Chapter 10.

1.2.3.1 Numeric Datatypes

1.2.3.1.1 Integer

Numeric data can be Integer, Float, Complex number or Boolean. In some variables, data is saved as a whole number. For example, "how many students are there in the class?" To answer such a question, an Integer-type variable is used. Integers are those datatypes which do not take any decimal places. That is, they take only the whole number, be it positive, negative or zero. For example, 8, −432 and 0 are integers. Let us consider the expression x = 3 + 5. Python calculates the right side of the equal sign, puts the result in a memory location and associates that location with the identifier x. In other words, it gives a name to that location and since the result of the expression on the right side is an Integer, the type of this variable will be *int*. In this case, we are having a variable and the identifier of this variable is x. It can also be said that the name of the variable is x. The value of the variable is 8 and the type of the variable is *int*. This can be verified from the command *print(type(x))* that returns *class int*, which means that its datatype is an Integer.

1.2.3.1.2 Float

Some variables have fractional values. If the question is "what is your height?", a common answer is of the form, say, 5.6 feet. It means we need a floating point representation to store the value of height. Floating point numbers are those numbers which consider decimal points also. For example, 43.9, −234.1, 00.4, etc. Let us take the expression y = 3/2. Python calculates the right side of the equal sign, puts the result in a memory location and associates that location with the identifier y. Since the result is a floating-point number, the type of this variable will be *Float*. The value of the variable is 1.5. This can be verified from the command *print(type(y))* that returns *class Float*, which means that its datatype is Float.

1.2.3.1.3 Complex

The Integer and the floating-point numbers are one-dimensional. Sometimes we need two-dimensional numbers also. To represent a two-dimensional number, we use another numeric datatype, complex number. The complex numbers are represented as an ordered pair *x+yj*, where x and y are numeric data. Here *x* is the real part and *y* is the imaginary part. For example, 9+8j or complex (9, 8) are complex numbers. Addition, subtraction, multiplication and division are the most common operations that we perform on complex datatypes.

1.2.3.1.4 Boolean

A binary variable that can have one of the two possible values – 0 (False or F) or 1 (True or T) is called Boolean. For example, when the question is "Are you happy?". The answer will be either "Yes" or "No". It means that a variable is needed which will store the Boolean datatype. AND, OR, XOR, etc. are the operations we perform with the Boolean datatype.

Now, let's look at the type bool. Let us consider an expression, z = False. Notice that in Python, both True and False are keywords and the first letter is in uppercase.

Make sure we do not type them as "true" or "false" (in lowercase). Python calculates the right side of the equal sign, puts the result in a memory location, associates that location with the identifier z, and since the result of the right side is a Boolean, the type is bool. Also notice that z was previously defined as "hello" and its type was a String. However, since the statement z = False, Python automatically puts the value False and associates it with z. So, the previous value of z which was "hello" is now overwritten and its new type would be bool. This is automatically done in Python and we don't have to worry about it.

There are a few more built-in datatypes such as String, List, Tuple, Dictionary and Set which come along with Python.

1.2.3.2 None Datatype

This is a special datatype that refers to *no value* or *null*. It does not mean *0* or *empty* or False or *undefined*. It is a real datatype as it can be assigned to a variable and occupies memory just like any other datatype. When a function does not have a return statement but we want to print the value returned by the function, this is handled by a *None* datatype.

1.2.3.3 Sequence Datatypes

1.2.3.3.1 String

To answer the question "what is your name?", the String type of variable is used. It is a sequence of characters, numbers and special characters. The String datatype is immutable. The operations that we perform with String datatype are concatenation, finding substring, etc. Let there be an assignment, z = "hello". Here again, Python calculates the right side, puts the result in a memory location, and associates that location with the identifier z, and since the right side is a group of characters, the type of this variable would be *str*, which stands for String. Just to recall, Python now has a variable with the identifier z, whose value is "hello", and its type is *str*.

1.2.3.3.2 List

This datatype is used to keep multiple data under one variable name. It is like an array in a programming language, for example, C. But C-Array is a homogeneous datatype, whereas a list need not be of the same datatype. In a list, the data is written within a pair of square braces [ ].

1.2.3.3.3 Tuple

This datatype is similar to the List datatype with the difference that once a tuple is defined, it contains a collection of unchangeable and indexed data. In a tuple, the data is written within a pair of round braces ( ).

1.2.3.4 Set Datatype

It is a collection of unordered and unindexed data. It is an immutable datatype which contains unique (no duplicate) data. In a set, the data is written within a pair of curly braces { }.

1.2.3.5 Mapping Datatype

1.2.3.5.1 Dictionaries

A dictionary is a special datatype where the data is kept in (key–value) pair. It is a collection which is unordered, changeable and indexed. This datatype works similar to a Dictionary in the real world. Keys of a Dictionary must be unique and of an immutable datatype such as Strings, Integers and Tuples, but the key–value can be repeated and of any type.

1.2.3.6 Array Datatype

This datatype holds data of multiple homogenous datatype together. It contains a collection of homogenous and indexed data. Array is not a core datatype. It means that in order to use an array, a separate package needs to be imported. Figure 1.1 shows the classification of datatypes.

1.3 CONSTANT

In any programming language, a constant is a value that does not change. Python does not have a constant but it has a *kind of constant*. It means Python does not

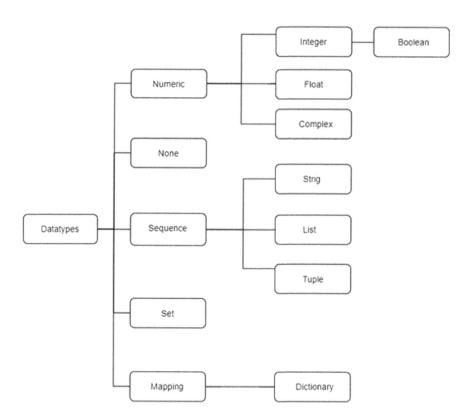

FIGURE 1.1 Classification of datatypes

support non-changing value once it is assigned to a variable but it has some standards to write the statement that feels like a constant. Constants in Python are defined and assigned a value in a module which can be imported to access it. A module is a file which contains variables and functions. We will discuss more about modules in Chapter 8. Usually, the user-defined constants are written all in uppercase. Some universal constants such as the value of *pi* that is used in connection to circles, the value of *tau* that returns the ratio of circumference to the radius of the circle and the Euler's number *e* are pre-defined in *math module*.

Example 1.6

```
import math
print("The value of PI is : ", math.pi)
print("The value of TAU is : ", math.tau)
print("The Euler's Number is : ", math.e)
```

The first line of the program imports *math module*. In order to access the constant like *pi, tau* and *e* defined in math module, we need to use *math.pi, math.tau and math.e*.

Output

```
The value of PI is : 3.141592653589793
The value of TAU is : 6.283185307179586
The Euler's Number is : 2.718281828459045
```

Let us discuss user-defined constants. We can write all the constants in a file with the *.py* extension and then import that file in the same way as we have imported math module in the above example.

Example 1.7

```
My_constant_file.py
# Create a separate file my_constant_file.py and write the
following two lines
Gravity = 9.8
Speed = 60
```

Here we have created a file *My_constant_file.py* wherein two constants, Gravity and Speed are defined.

```
import my_constant_file
print("The Gravity of Earth is ",my_constant_file.Gravity)
print("The Speed of Earth is ",my_constant_file.Speed)
```

Observe that we have imported the *My_constant_file* and then used the constants as we have used in the previous example.

Output

```
The Gravity of Earth is 9.8
The Speed of Earth is 60
```

1.4 STATEMENT AND EXPRESSION

A statement is a syntactic unit of any programming language that instructs some activity to be completed. A program may contain a sequence of one or more statements. An expression is a part of code that calculates a value. For example, when we write x = 9, it is an expression as it evaluates a value which is 9. Similarly, when we write 3 × 4, it is an expression because it calculates the value 12. Here, 3 by itself is also an expression, which is usually called a sub-expression and 4 is also a sub-expression.

"Hey!" is also an expression, because it evaluates to a String, where the value of the String is "Hey!"; "True" is also an expression because it is a Python keyword and it evaluates to the value of true, which is a Boolean type. A statement is an instruction that Python can execute. Any instruction that we write in Python and that Python can execute is a statement.

1.5 INPUT STATEMENTS

To get an input from the keyboard, the *input()* function is called. The general format of the input function is

> *Variable_name = input(["Prompt Message"])*

The prompt message (optional) is used to show a message to the user before he types anything. It is basically used for better understanding of the program.

Write a program to get input from the user. Let's see what happens when we run the program in the Python shell.

Example 1.8

```
>>> Age = input("Enter Your Age ")
Enter Your Age
```

Observe that the prompt appears in the Python shell. Now, Python is waiting for the user to type something. Such a type of programming style is known as *interactive mode*.

Output

```
>>> Age = input("Enter Your Age ")
Enter Your Age 23
>>>
```

We can write the program in *script mode* also. We need to create a new file by clicking on Menu > File > New File; write the code and save it with a file name with *.py* extension.

Example 1.9

```
name = input("Enter Your Name")
age = input("Enter Your age")
print("You are ",name,"and your age is ",age)
```

Output

```
Enter Your Name Rama
Enter Your age 23
You are Rama and your age is 23
>>>
```

Example 1.10

```
no1 = input("Enter Number 1 : ")
no2 = input("Enter Number 2 : ")
sum = no1 + no2
print("The Sum is", sum)
```

Output

```
Enter Number 1 : 5
Enter Number 2 : 8
The Sum is 58
>>>
```

The sum of 5 and 8 is 13, but here it is 58. An ERROR...? No, it is because when we get an input from the keyboard by using input(), it takes strings. When we try to do mathematical operations on a character String, then we will get an error. But in this case, we get 58 as a result of the String concatenation operation. If we use mathematical operations like subtraction or division (– or /) we will get error messages.

Example 1.11

```
no1 = input("Enter Number 1 : ")
no2 = input("Enter Number 2 : ")
sum = no1 / no2
print("The Division is ",sum)
```

Observe the error message "unsupported operand types for /". The two operands we have taken are strings (not numbers). We cannot perform division operations with String datatype.

Output

```
Enter Number 1 : 22
Enter Number 2 : 11
Traceback (most recent call last):
  File "C:\Users\RakeshN\AppData\Local\Programs\Python\Py
  thon37-32\chap-2.py", line 68, in <module>
    sum = no1 / no2
TypeError: unsupported operand type(s) for /: 'str' and 'str'
>>>
```

To overcome this problem, type casting is to be done. It means the input value needs to be changed from String to Integer or Float datatype.

Example 1.12

```
num1 = int(input("Enter Number 1 : "))
num2 = int(input("Enter Number 2 : "))
sum = num1 + num2
print("The Sum is ",sum)
```

Observe that in line 1 and line 2, type casting is done with Integer datatype. It means the input that we got in String datatype is converted to Integer before executing the third line. Now we get the sum of 5 and 8 as 13, which is correct.

Output

```
Enter Number 1 : 8
Enter Number 2 : 5
The Sum is 13
>>>
```

The eval() function

When the user enters some characters from the keyboard using an input statement, it produces a String. If the user wishes to treat that input as a number, then type casting is necessarily required for the conversion. Depending upon the user input, the datatype of the variable can be decided Python provides the *eval()* function that attempts to evaluate a String in the same way that the interactive shell would evaluate it.

Example 1.13

```
x = eval(input("Enter an Integer :"))
print("The Entered Integer is ",x," Datatype is ",type(x))
y = eval(input("Enter a Floating Point Number :"))
print("The Entered Floating Point Number is ",y," Datatype
is ",type(y))
```

Output

```
Enter an Integer :5
The Entered Integer is 5 Datatype is <class 'int'>
Enter a Floating Point Number :5.5
The Entered Floating Point Number is 5.5 Datatype is <class
'Float'>
>>>
```

The *eval()* function dynamically translates the text provided by the user into an executable form that the program can process. This allows the user to provide input in a variety of flexible ways. The *"eval()"* does not work with String datatype. Using the *eval()* function, the user can enter values assigned to multiple variables at the same time. This is a unique feature in Python. On the left-hand side of any assignment, we can have multiple variables separated by a comma.

Example 1.14

```
x, y = eval(input("Enter two numbers separated by a comma : "))
print("The Sum is ",x+y)
```

Output

```
Enter two numbers separated by a comma : 12, 4.9
The Sum is 16.9
>>>
```

1.6 OUTPUT STATEMENTS

Output statements are the statements that are used to show the result on the screen. For that purpose, we use the *print()* statement.

A simple print statement looks like

print("objects",[sep = '', end = '\n', file = sys.stdout, flush = false])

All non-keyword arguments are converted to strings like *str()* does, written to the stream, separated by comma *"sep"* and followed by *"end"*. Both *sep* and *end* must be strings. The use of *sep* and *end* are optional. If no objects are given, *print()* command will print a blank line. The file argument must be an object with a *write(string)* method; if it is not present or *None* datatype, *sys.stdout* will be used. Since printed arguments are converted to text strings, *print()* cannot be used with binary mode file objects. For binary mode file objects, we use *file.write(...)* instead.

Whether the output is buffered or not is usually determined by *file*, but if the flush keyword argument is true, the stream is forcibly flushed. Observe that the *print()* statement consists of the following parts:

- The word print.
- An opening parenthesis.
- An opening quotation mark.
- A statement to be printed.
- A closing quotation mark.
- A closing parenthesis.

Each of the above plays a very important role in the code.

A *print()* function takes its arguments (it may accept more than one argument and may also accept no argument), converts them into human-readable form if needed, and sends the resulting data to the output device (usually the console). In other words, anything we put into the *print()* function will appear on our screen; *print()* does not evaluate anything.

The name chosen while defining the function should be self-explanatory. In other words, it must highlight the task to be done by the function, for example, the name of the print function is self-evident.

Python functions are more versatile. Depending on the individual needs, they may accept any number of arguments. Some Python functions don't need any argument. Despite the number of needed/provided arguments, Python functions strongly demand the presence of a pair of parentheses – opening and closing ones, respectively. If we want to deliver one or more arguments to a function, we place them inside the parentheses. If we want to use a function that doesn't take any argument, we still have to have the parentheses.

Note that in order to distinguish ordinary words from function names, a pair of empty parentheses are placed after the function names, even if the corresponding function wants one or more arguments. This is a standard convention.

Example 1.15

```
>>> print("I am Learning Python")
I am Learning Python
>>>
```

In this example, the only argument delivered to the *print()* function is a String. As it can be seen, the String is delimited with quotes. Almost anything we put inside the quotes will be taken literally, not as code, but as data. The function name (print, in this case) along with the parentheses and argument(s), forms the function invocation. When Python encounters an invocation like this one, it:

- First checks if the specified name is valid.
- Checks if the function's requirements for the number of arguments allow it to invoke the function.
- Leaves the code for a moment and jumps into the function we want to invoke; of course, it takes the argument(s) too and passes them to the function.
- Executes the function's code, causes the desired effect, evaluates the desired result(s) (if any), and finishes its task.

- Finally, returns to the code (to the place just after the invocation) and resumes its execution.

A print statement can be empty (i.e., it may contain no instruction at all) but it must not contain two, three or more instructions. This is strictly prohibited. Python makes one exception to this rule – it allows one instruction to spread across more than one line (which may be helpful when our code contains complex constructions).

Example 1.16

```
print("I am learning Python")
print("I am enjoying Learning it")
```

The program invokes the *print()* function twice, and two separate lines are observed in the console – this means that *print()* begins its output from a new line each time it starts its execution; this behavior can be changed but can also be used for user's advantage. Each *print()* invocation contains a different String as its argument and the console content reflects it – this means that the instructions in the code are executed in the same order in which they have been placed in the source file; no next instruction is executed until the previous one is completed (there are some exceptions to this rule, which we can ignore for now).

Example 1.17

```
print("I am learning Python")
print()
print("I am enjoying Learning it")
```

The empty *print()* invocation is not as empty as we may have expected – it gives output of an empty line. This is not the only way to produce a newline in the output console.

1.7   FORMATTING OUTPUT STATEMENTS

Formatting refers to the presentation or layout of a text. When we want to control the way the output should be displayed, we use formatted outputs. There are several ways we format any output statement in Python.

1.7.1   Escape Character

The backslash (\) has a very special meaning when used inside the strings – this is called the *escape* character. In other words, the backslash doesn't mean anything in itself, but is only a kind of announcement that the next character after the backslash has a different meaning too. The word *escape* should be understood specifically. It means that the series of characters in the String escapes for the moment (a very short moment) to introduce a special inclusion. The letter *n* placed after the backslash comes from the word *newline*.

TABLE 1.1

Escape Sequence

Escape Character	Meaning
\n (New line)	Use it to shift the cursor control to the new line.
\t (Horizontal tab)	Use it to shift the cursor to a couple of spaces to the right in the same line.
\' (Apostrophe or single quotation mark)	Use it to display the single-quotation mark.
\" (Double-quotation mark)	Use it to display the double-quotation mark.
\\ (Backslash)	Use it to display the backslash character.

Example 1.18

```
>>> print("I am learning Python \n and I enjoy learning
it")
I am learning Python
and I enjoy learning it
>>>
```

Both the backslash and the *n* form a special symbol named a newline character, which urges the console to start a new output line. Some more escape sequences in Python are shown in table 1.1.

Example 1.19

```
print("I am learning Python \n and I am enjoying it")
print("I am learning Python \t and I am enjoying it")
print("I\'m learning Python and I\'m enjoying it")
print("I am learning \"Python\" and I am enjoying it")
print("I am learning Python \\ I am enjoying it")
```

Output

```
I am learning Python
and I am enjoying it
I am learning Python and I am enjoying it
I'm learning Python and I'm enjoying it
I am learning "Python" and I am enjoying it
I am learning Python \ I am enjoying it
>>>
```

There are many more escape sequences which we'll discuss in due course of time. Look at the next few examples.

Example 1.20

```
print("I am learning Python; I am enjoying it")
```

Output:

```
I am learning Python; I am enjoying it
>>>
```

Example 1.21

```
print("I am", "learning Python", "and", "I am enjoying it")
```

Observe the *print()* statement; it contains four arguments. All of them are strings. The arguments are separated by commas. In this case, the commas separating the arguments play a completely different role than the comma inside the String. The latter is a part of Python's syntax whereas the former one is intended to be shown in the console.

Output

```
I am learning Python and I am enjoying it
>>>
```

The way in which the arguments are passed into the *print()* function is the most common in Python, and is called the *positional argument* (this name comes from the fact that the meaning of the argument is dictated by its position). Python offers another mechanism for *passing the arguments* which can be helpful when we want to convince the *print()* function to change its behavior a bit. The mechanism is called *keyword arguments*. The name stems from the fact that the meaning of these arguments is taken not from its location (position) but from the special word (keyword) used to identify them.

1.7.2   SEP AND END

The *print()* function has two keyword arguments that can be used. First one is named *end*. It is necessary to know some rules in order to use it. They are:

- a keyword argument consists of three elements: a keyword identifying the argument (end here), an equal sign (=) and a value assigned to that argument;
- any keyword argument must be placed after the last positional argument (this is very important).

The default behavior reflects the situation where the *end* keyword argument is implicitly used in the following way:

Example 1.22

```
print("I am learning Python")
print("I am enjoying it.")
```

Output

```
I am learning Python
I am enjoying it.
>>>
```

The String assigned to the *end* keyword argument can be of any length (even length 0). The *end* keyword argument determines the characters the *print()* function sends to the output once it reaches the end of its *positional arguments*.

We have made use of the *end* keyword argument and set it to a String containing one space in the example below.

Example 1.23

```
print("I am learning Python", end = "")
print("I am enjoying it.")
```

As the *end* argument has been set to nothing, the print() function outputs nothing too, once its positional arguments have been exhausted. No newlines have been sent to the output.

Output

```
I am learning PythonI am enjoying it.
>>>
```

Example 1.24

```
print("I am learning Python", end = "\n")
print("I am enjoying it.")
```

Observe that the end argument is set to \n. This enables a new line inserted at the end of the first print statement.

Output

```
I am learning Python
I am enjoying it.
>>>
```

Example 1.25

```
print("I am learning Python", end = "")
print("I am enjoying it", end = "...")
```

In the first line, 'end = "" ', indicates that instead of the end-of-line, an empty String is printed. As a result, the output of the second line combines with the first line. In the second line, 'end = "..." ', makes three dots printed at the end.

Output

```
I am learning PythonI am enjoying it...
>>>
```

Generally, the *print()* function separates its arguments with spaces. This behavior can be changed too with the help of the second keyword argument that is named *sep* (derived from separator). Note that the argument's value may be an empty String also.

Example 1.26

```
print("I am learning Python","and","I am enjoying it.",sep = "_")
print("I am learning Python","and","I am enjoying it.",sep = "...")
print("I am learning Python","and","I am enjoying it.",sep = "")
print("I am learning Python","and","I am enjoying it.",sep = "\n")
```

The first *print()* function uses 'sep="_" '. It can be seen from the output below that all the parameters in this print statement are separated by an underscore. Similarly in the second print statement the keyword argument is 'sep="..." '. This makes all the printable arguments separated by "...".

Output

```
I am learning Python_and_I am enjoying it.
I am learning Python...and...I am enjoying it.
I am learning PythonandI am enjoying it.
I am learning Python
and
I am enjoying it.
>>>
```

Both keyword arguments (sep and end) may also be used in one invocation.

Example 1.27

```
print("I am learning Python","and","I am enjoying it.",sep = "...",end = "***")
```

Observe the output after every argument in the print statement; three dots (...) are printed and at the end three stars (***) are printed.

Output

```
I am learning Python...and...I am enjoying it.***
>>>
```

1.7.3   String Formatting with format()

String formatting is yet another kind of statement which can be used in input/output statements that allow multiple substitutions and value formatting. Positional formatting is used in this method. The statement "*format()*" takes two different forms of parameters:

Positional parameters – List of parameters that may be accessed with an index of parameters within a pair of curly braces { }.
Keyword parameters – List of parameters of sort key value that may be accessed with a key of parameters within a pair of curly braces { }.

The format() reads the type of arguments passed to it and presents it according to the format codes defined in the String. The basic form is:

print(string.format(value))

The value can be any datatype including a variable.

Example 1.28

```
print("I am learning {} and I am enjoying it
".format("Python"))
```

In this example, the value that is passed is "Python".

Output

```
I am learning Python and I am enjoying it
>>>
```

It can be observed from the output that in place of the pair of empty curly braces, Python is substituted.

Example 1.29

```
str = "Python"
print("I am learning {} and I am enjoying it ".format(str))
```

In this example, the output is the same as the previous example but the parameter that is passed is a variable.

Output

```
I am learning Python and I am enjoying it
>>>
```

Example 1.30

```
print("I am learning {0} and I am {1} it ".format("Python",
"Enjoying"))
```

In this example, there are two parameters passed in the format(). In order to define where to substitute which parameter, an ordering is required. The String "Python" appears as the first parameter and 0 is assigned to it. Similarly, as the String "Enjoying" appears second, 1 is assigned to it. Based on the values appearing inside the curly braces, the respective String is substituted.

Output

```
I am learning Python and I am Enjoying it
>>>
```

Example 1.31

```
str1 = "Python"
str2 = "Enjoying"
print("I am learning {0} and I am {1} it ".format(str1,str2))
```

The output is the same as the output of the previous example. In this example, instead of an exact String, variables are passed as parameters.

Output:

```
I am learning Python and I am Enjoying it
>>>
```

Example 1.32

```
name = input("Enter Your Name : ")
age = int(input("{0}, Please Enter your age :
".format(name)))
print("{0}, you are {1} years old".format(name, age))
```

Output

```
Enter Your Name : Rama
Rama, Please Enter your age : 23
Rama, you are 23 years old
>>>
```

1.7.4 NUMBERS FORMATTING WITH FORMAT()

Like String formatting, we can also format numbers using the format specifier. The format specification for printing an Integer number is %wd, where % is a symbol used to inform the interpreter about where the formatting is to be done, w specifies the minimum field width for the output and d specifies the value to be printed as an Integer. If a number is less than the specified width, leading blanks will appear as necessary. The number is written *right justified* in the given field width. "+d" with a positive number prints "+" sign before the number and "–d" with a negative number prints "–" sign before the number, "–d" with a positive number prints no sign before the number whereas "–d" with a negative number prints "–" sign before the number.

Example 1.33

```
print("{:d}".format(234))
print("{:6d}".format(234))
print("{:08d}".format(234))
print("{:+d}".format(234))
print("{:+d}".format(-234))
print("{:-d}".format(234))
print("{:-d}".format(-234))
```

Observe the second print statement {:6d} in the print command. As specified, the field length is 6. Out of which three are occupied by the data 234 and as the numbers are right justified, the rest three places to the left are filled with blank spaces.

Output

```
234
234
00000234
+234
-234
234
-234
>>>
```

The format specification for printing a floating-point number is %w.pf, where w specifies the width with p decimal places for the output and f specifies that the value to be printed is a floating-point number. If a number is less than the specified width, leading blanks will appear as necessary. The number is written *right justified* in the given field width. As in "+d", with "+f", "+" sign is printed before a positive floating-point number; with "-f", a "–" sign is printed before the floating-point number. Similar to "–d", "–f" with a positive floating-point number prints no sign before the number and "–f" with a negative floating-point number prints – sign before the number.

Example 1.34

```
print("{:f}".format(234))
print("{:2.2f}".format(234))
print("{:8.1f}".format(234))
print("{:+f}".format(234))
print("{:+f}".format(-234))
print("{:-f}".format(234))
print("{:-f}".format(-234))
```

Output

```
234.000000
234.00
   234.0
+234.000000
-234.000000
234.000000
-234.000000
>>>
```

The format specification for printing a String is *%w.ps*, where *w* specifies the width with first *p* characters to be displayed in the output and *s* specifies that the value to be printed is a String. If the String length is less than the specified width, trailing blanks will appear as necessary. The String is written *left justified* in the given field width.

Example 1.35

```
print("{:s}".format("Python"),end = "*\n")
print("{:20s}".format("Python"),end = "*\n")
print("{:8.2s}".format("Python"))
```

Observe that in the first print command, we have used end="*\n". It means "*" is printed and the cursor moves to the next line. In the second print command {:20s} indicates that the field width is 20. As strings are printed left justified, first the String "Python" is printed and the rest of the places to the right are filled with blanks. In this statement we also have "end="*\n"", which indicates that a "*" is printed at the end of specified field width and then carriage return. In the third print command {:8.2s} indicates that only the first two characters are printed.

Output

```
Python*
Python              *
Py
>>>
```

There are many such format strings available as listed in table 1.2.

TABLE 1.2
Format Strings

Type	Meaning
D	Decimal Integer
C	Corresponding Unicode character
B	Binary format
O	Octal format
X	Hexadecimal format (lower case)
X	Hexadecimal format (upper case)
N	Same as "d". Except it uses current locale setting for number separator
E	Exponential notation (Lowercase e)
E	Exponential notation (uppercase E)
F	Displays fixed point number (Default: 6)
F	Same as "f". Except displays "inf" as "INF" and "nan" as "NAN"
G	General format. Rounds number to "p" significant digits. (Default precision: 6)
G	Same as "g". Except that it switches to "E" if the number is large.
%	Percentage. Multiples by 100 and puts % at the end.

Apart from these format strings, there are a few other symbols used for alignment when assigned a certain width. They are listed in table 1.3.

TABLE 1.3
Symbols Used for Alignment

Type	Meaning
<	Left aligned to the remaining space
^	Center aligned to the remaining space
>	Right aligned to the remaining space
=	Forces the signed (+) (−) to the leftmost position

Example 1.36

```
print("{:d}".format(234))
print("{:*<20d}".format(234))
print("{:*>20d}".format(234))
print("{:*^20d}".format(234))
print("{:=20d}".format(+234))
print("{:=20d}".format(-234))
```

In the second print statement, we have "{*<20d}". This indicates the field width is 20. The symbol "<" indicates left justified and "*" indicates that except for the data, the rest of the 17 spaces are filled with "*". The third and the fourth print statements are right justified (>) and center justified (^) respectively. In the fifth and the sixth statements the format specified is "{:=20d}". It is the same as using "{:20d}" as discussed earlier in {:6d} in example 1.33.

Output

```
234
234*****************
****************234
********234**********
           234
-          234
>>>
```

1.8   COMMENT STATEMENT

To comment a line, or a portion of a line, use the # (sharp or hash symbol). All the characters after this symbol are ignored by Python. Any line is completely ignored by Python if it starts with a # symbol.

Example 1.37

```
# To Calculate the Perimeter of a Rectangle
Rectangle_Length = 5 # Length of the Rectangle
Rectangle_Breadth = 10 # Breadth of the Rectangle
Rectangle_Perimeter = 2 * ( Rectangle_Length + Rectangle_
Breadth )
print("The Perimeter of rectangle is : ",Rectangle_Perimeter)
```

The first line is completely ignored by Python because it starts with a # symbol. Notice that if we have more than one line of comment, each of the lines have to start with #. We can comment on a portion of a line also. Observe the second and the third statement in the program. A portion of the program is commented. All the characters after # are treated as a comment. Triple quotation mark (""" comment """) is used to comment multiple lines at a time.

Example 1.38

```
"""
This is a Demonstration
to Comment
multiple lines
"""
# To Calculate the Perimeter of a Rectangle
Rectangle_Length = 5 # Length of the Rectangle
Rectangle_Breadth = 10 # Breadth of the Rectangle
```

```
Rectangle_Perimeter = 2 * ( Rectangle_Length + Rectangle_
Breadth )
print ("The Perimeter of rectangle is : ",Rectangle_Perimeter)
```

MULTIPLE CHOICE QUESTIONS

1. Which of the following is incorrect?
   a) An identifier is the name given to a variable, a function or an object
   b) In Python different datatypes can be stored in a variable
   c) A variable is the name given to an identifier
   d) Variable and identifier are one and the same

2. Which of the following is incorrect?
   a) A variable has properties such as value, type and scope
   b) Variables and identifiers are the same
   c) A variable is a reserved location in a computer's memory to store data
   d) A variable is the name given to an identifier

3. For the statement x = 5 + 9 , what is the type?
   a) Integer
   b) String
   c) Float
   d) Boolean

4. For the statement x = 32 / 8, what is the type?
   a) Integer
   b) String
   c) Float
   d) Boolean

5. For the statement x = input("What is your age"), what is the type?
   a) Integer
   b) String
   c) Float
   d) Boolean

6. For the statement x = int(input("What is your age")), what is the type?
   a) Integer
   b) String
   c) Float
   d) Boolean

7. Find the output of the statement x = (int(4.2)/2)
   a) 0.2
   b) 2
   c) 2.0
   d) None of the above

8. Find the output of the statement
   x = 10
   y = 20
   print("x is ",y," y is ",x)
   a) x is 10 y is 20
   b) x is 20 y is 10
   c) x is 20 y is 20
   d) x is 10 y is 10

9. Which of the following is an incorrect variable name?
   a) Ab_c
   b) Ab@c
   c) Abc
   d) _Abc

10. Which of the following is an incorrect statement?
   a) Ab9c = 1
   b) @Abc = 1
   c) 9Abc = 1
   d) Ab–c = 10

11. Which of the following is an invalid statement?
   a) _A = 11
   b) __A = 1
   c) __A__ = 1
   d) None of the above

12. Is Python case sensitive?
   a) Machine dependent
   b) Yes
   c) No
   d) Maybe

13. Maximum possible variable length in Python is
   a) 32 Characters
   b) 64 Characters
   c) 128 Characters
   d) None of the above

14. Which is an invalid statement?
   a) xyz=100,00,000
   b) x y z = 100 200 300
   c) x_y_z=100,00,000
   d) x,y,z=100,200,300

15. If key and value pair is needed, which datatype is to be used?
   a) int
   b) list
   c) tuple
   d) dictionary

16. If how-are-you is to be printed, the print statement will be?
    a) print("how-","are","-you")
    b) print("how","are","you",sep="-")
    c) print("how"+"-"+"are"+"-"+"you")
    d) All of the above

17. Which one is core Python datatype?
    a) class
    b) int
    c) String
    d) Dictionary

18. What will be the output when x = y is executed?
    a) y is assigned to x
    b) x is assigned to y
    c) check if x is equal to y
    d) Error

19. What will be the output of 1 + 2==3?
    a) True
    b) False
    c) Machine dependent
    d) Error

20. Complex numbers are represented as
    a) 5 + 6j
    b) 5 + 6i
    c) Complex(5,6)
    d) 5 + 6J

DESCRIPTIVE QUESTIONS

1. Write the difference between variables and identifiers.
2. Describe core datatypes in Python.
3. Write the rules for naming a variable in Python.
4. Write a note on eval() function.
5. List different escape sequences.

PROGRAMMING QUESTIONS

1. Write a Python program that calculates the area and perimeter of a circle.
2. Write a Python program that converts your age to the number of days. Assume that there are 365 days in a year. If you are 18 years old, the output should be 6570.
3. Write a Python program that asks the user for an Integer "x" and prints the value of y after evaluating the expression: $x^4 + x^{123}$.

4. Write a Python program that finds the final amount by simple interest, if principal, rate of interest and time are given.
5. Write a Python program that converts an Integer to a binary number (use String conversion).

ANSWER TO MULTIPLE CHOICE QUESTIONS

1.	C	2.	B	3.	A	4.	C	5.	B
6.	A	7.	C	8.	B	9.	B	10.	A
11.	D	12.	B	13.	D	14.	B	15.	D
16.	D	17.	A	18.	D	19.	A	20.	B

2 Operators

LEARNING OBJECTIVES

After studying this chapter, the reader will be able to:

- Understand and identify different types of operators used in Python
- Comprehend the order of Precedence of operators
- Use type casting

2.1 INTRODUCTION

All mathematical operations consist of operands and operators. Operands can be variables or constants. Operations act upon operands. Arithmetic operators are one among many different types of operators used in Python. In this chapter, we are going to discuss all the operators that are frequently used in Python.

2.2 TYPES OF OPERATORS

Operators are the symbols giving commands to the Python interpreter to perform mathematical or logical operations. Some operators act as unary operators, some as binary operators and some operators act as unary or binary depending on how the expression is written. Python operators can be classified into several categories. They include:

- a. Arithmetic operators.
- b. Relational operators.
- c. Logical operators.
- d. Increment and decrement operators.
- e. Bitwise operators.
- f. Special operators.

2.2.1 ASSIGNMENT OPERATOR

An assignment operator is a statement which assigns the right-side value to the identifier which is on the left of the "=" sign. The *equal sign* "=" is used to assign a value to a variable. In other words, an equal sign means an association of an identifier with a variable. The expression $x = 2$ does not mean that x is equal to 2. It simply means to take the value of the right side, which in this case is 2, and store it in a location in the memory of the computer which is identified by x. It is best if we think of $x = 2$

DOI: 10.1201/9781003219125-3

as x is associated with the memory location which holds the value of 2. The simplest form of this statement is ***Variable_name = Expresssion***

Example 2.1

```
>>> x = 4 * 6
>>> print(x)
24
>>>
```

Here $x = 4 * 6$ is a statement. Remember that the right side of this statement is an expression but the entire $x = 4 * 6$ is an assignment statement. Similarly, $A = 4 + 8$ is an assignment because it associates the value of the expression on the right of the equal sign with the identifier A. This is also a statement. It associates the value of the expression on the right side of the equal sign with the identifier.

In mathematics, if we write $a = a + 5$, it means to *find the memory location which is identified by a, take the value of the location and add 5 to it and re-assign the variable a.* There must be a pre-existing value of "*a*" before writing such a statement. Note that it will be the same memory location and this operation overrides the previous value of the location x. This statement can be written as $a += 5$.

Example 2.2

```
>>> a = 5
>>> a += 5
>>> print("The value of a is ",a)
The value of a is 10
>>>
```

Unlike any other programming language, Python assigns more than one variable to the respective values in a single statement. It means there can be more than one variable on the left of an assignment statement. In other words, multiple assignment statements can be issued over a single statement.

Example 2.3

```
>>> x, y = 5, 8
>>> print("x = ",x,"y = ",y)
x = 5 y = 8
>>>
```

As Python assigns values to more than one variable using a single "=" statement, this symbol is also used for exchanging data (swapping) in a single statement.

Example 2.4

```
>>> x1 = 7
>>> x2 = 5
>>> x3 = 6
>>> x4 = 4
>>> print("Data Before Swapping", x1, x2, x3, x4)
Data Before Swapping 7 5 6 4
>>> x2, x3 = x3, x2
>>> print("Data after Swapping", x1, x2, x3, x4)
Data after Swapping 7 6 5 4
>>>
```

In the above example, observe that there are four variables, x1, x2, x3 and x4 with values 7, 5, 6 and 4 respectively. We exchange the values of variable x2 and x3 by using the statement x2, x3 = x3, x2 a single assignment statement. It is the same as writing temp = x2, x2 = x3 and x3 = temp in a programming language like C. But in Python it can be done in a single statement.

2.2.2 ARITHMETIC OPERATORS

Python provides all basic arithmetic operators. They are listed in table 2.1. The symbols which are used for mathematical operations are +, −, *, /, **, %, //, and a unary minus sign.

Example 2.5

```
>>> 9 * 8
72
>>>
```

In Python, when the expression 9 * 8 is written, (the asterisk is the multiplication symbol) it means the Python is asked to multiply 9 and 8.

TABLE 2.1
Arithmetic Operators

Operator	Meaning
+	Addition
−	Subtraction and unary minus
*	Multiplication
/	Division
**	Exponential
%	Modulo Davison
//	Integer Division

Example 2.6

```
>>> -505
-505
>>>
```

If the expression is negative 505, it means, the negative operator acted on the operand 505. Here the minus sign is called a unary operator as it works on a single operand.

Example 2.7

```
>>> 52 - 17
35
>>>
```

In this example, the minus sign is a binary operator because it acts on two operands.

Example 2.8

```
>>> -52 - 17
-69
>>>
```

However, in this example, the negative sign in (–52) is a unary operator, because it only acts on 52 and the negative sign between the two operands is a binary operator as it acts on two operands. When both operands on single arithmetic expressions are integers, the expression is called Integer expression. Integer expressions always yield an Integer value on addition, subtraction, remainder and Integer division.

Example 2.9

```
>>> a = 43
>>> b = 3
>>> a + b
46
>>> a - b
40
>>> a * b
129
>>> a % b
1
>>> a // b
14
>>>
```

An * (asterisk) sign is a multiplication operator.

```
>>> 5 * 3
15
>>> 5. * 3
15.0
>>> 5 * 3.
15.0
>>> 5. * 3.
15.0
>>>
```

The rules for * operations are:

- When both operands are integers, the result is an Integer.
- When at least one operand is a Float, the result is a Float.

The result produced by the division operator is always a Float regardless of whether the operand is floating-point or not.

Example 2.10

```
>>> 3 / 6.
0.5
>>> 6. / 3
2.0
>>>
```

A ** (double asterisk) sign is the exponentiation (power) operator. Its left argument is the base and right, the exponent. The mathematical expression 5^3 is written as 5 ** 3 in Python.

Example 2.11

```
>>> 5 ** 3
125
>>> 5. ** 3
125.0
>>> 5 ** 3.
125.0
>>> 5. ** 3.
125.0
>>>
```

The rules for ** operations are

- When both ** operands are integers, the result is an Integer.
- When at least one ** operand is a Float, the result is a floating-point number.

The modulus operator shows the remainder. When we write x % n, it refers to the remainder when x is divided by n. During modulo division, the sign of the result is always the same as the sign of the second operand. In the expression, the second operand is n. The result we obtain may be any number in the range $[-(n-1),\ldots,0,\ldots, (n-1)]$. When the second operand is positive, to find the modulus operation, go around the circle in a clockwise direction. When the second operand is negative, to find the modulus operation, go around the circle in an anti-clockwise direction as shown in figure 2.1. The pairs (–1, 3), (–2, 2), (–3, 1) are called modular inverse. As $-(-1) + 3 = -(-2) + 2 = -(-3) + 1 = 4$.

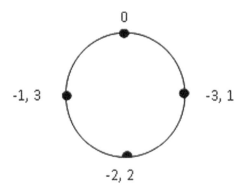

FIGURE 2.1   Modulus Operation

Example 2.12

```
>>> a = -15
>>> b = 4
>>> a % b
1
>>>
```

Example 2.13

```
>>> a = 15
>>> b = -4
>>> a % b
-1
>>>
```

Example 2.14

```
>>> a = -15
>>> b = -4
>>> a % b
-3
>>>
```

Observe that in example 2.12, as the second operand (b = 4) is positive, the remainder is positive. In example 2.13 and example 2.14, the second operand (b = −4) is negative and the remainder is negative. Based on the sign of the second operand (in this case either b = 4 or b = −4), the sign of the modulo operation can be negative or positive. When the second operand is negative (say −n), the result can be any number in the range [−(n−1), −(n−2), ..., −3, −2, −1,0] and when the second operand is positive (say n), the result can be any number in the range [0, 1, 2, 3, ..., (n−2), (n−1)]. In all other programming languages for modulus operation, the result is in the range [0, 1, 2, 3, ..., (n−1)]. When the mod operation is giving a negative sign, subtract the second operand to it to get a positive remainder.

A // (double slash) sign is an Integer divisional operator. It differs from the standard/operator in two ways:

• The result doesn't have a fractional part, in other words, the results are always an Integer.
• It confirms to the Integer vs Float rule. It means when both operands are an Integer then the output is an Integer whereas if at least one operand is a floating point, the result is a floating point.

Example 2.15

```
>>> 6 // 5
1
>>> 6. // 5
1.0
>>> 6 // 5.
1.0
>>> 6. // 5.
1.0
>>>
```

The result of the Integer division is always rounded to the nearest Integer value that is less than the real (not rounded) result. This is very important – rounding always goes to the lesser Integer.

2.2.2.1 Precedence of Arithmetic Operators

Evaluation of expression is based on the precedence of operators. If we want to evaluate 3 − 6*7, whether the meaning of this expression is (3−6) * 7 or 3 − (6 * 7)?

TABLE 2.2

Operators

Operator	Meaning
+, −	unary plus and unary minus
**	Exponentiation
*,/, %, //	Multiply, divide, modulo and whole (floor) division
+, −	Addition and Subtraction
= ,%= ,/= ,//= ,-= ,+= ,*=, **=	Assignment operators

Example 2.16

```
>>> 3 - 6 * 7
-39
>>>
```

Here, if the subtraction is performed first and then the multiplication, the result would be different. The rules which are followed by Python to determine in which order the operations should be performed are called *precedence rules*. Table 2.2 shows the operators with their meaning.

In the 1st row we have +, − and in the 4th row we also have + and −. The difference is, that the + and − in the 4th row are binary operations, whereas the + and − in the 1st row is a unary operation. According to the precedence rule given in table 2.2, in example 2.16 Python performs multiplication first and then performs the subtraction as a result −39. Parentheses are always used to explicitly show the user's point of view. If correct parentheses are used, then the user doesn't have to look after precedence rules. The binding of the operator determines the order of computations performed by some operators with equal priority, put side by side in one expression. Most of the operators in Python have left-sided binding, which means that the calculation of the expression is conducted from left to right.

Example 2.17

```
>>> 16 % 6 % 3
1
>>>
```

There are two possible ways of evaluating this expression. They are:

- From left to right: first 16 % 6 gives 4, and then 4 % 3 gives 1.
- From right to left: first 6 % 3 gives 0, and then 16 % 0 causes a fatal error.

Now try 4 ** 3 ** 2

The two possible ways of evaluating the expression are:

* From left to right: first 4 ** 3 = 64; and then 64 ** 2 = 4096.
* From right to left: first 3 ** 2 = 9; and then 4 ** 9 = 262144.

Example 2.18

```
>>> a = 4 ** 3 ** 2
>>> print(a)
262144
>>>
```

The result clearly shows that the exponentiation (**) operator uses right-sided binding.

2.2.3 RELATIONAL OPERATORS

These operators compare two operands. For example, if something is greater than something else, or less than, or equal to, or not equal to, or greater than or equal to, or less than or equal to. While comparing operators, we get Boolean answers. Table 2.3 shows commonly used relational operators.

Start with a very simple example 10 > 5.

Example 2.19

```
>>> a = 10
>>> b = 5
>>> c = a > b
>>> print(c)
True
>>> d = a < c
>>> print(d)
False
>>>
```

As 10 is greater than 5, the first print(c) gives True whereas the second print(d) gives False.

One of the mistakes that programmers commit is to use one = sign rather than two. Don't forget this important distinction.

1. = is an assignment operator (a = b assigns a with the value of b).
2. == is a question that are these values equal? (a == b compares a and b).

It is a binary operator with left-sided binding. It needs two arguments and checks if they are equal.

TABLE 2.3
Relational Operators

Operator	Meaning	Usage
>	Greater than	x > y
<	Less than	x < y
==	Equal to	x == y
!=	Not equal to	x != y
>=	Greater than or equal to	x >= y
<=	Less than or equal to	x <= y

Example 2.20

```
>>> a = 15
>>> b = 10
>>> print(a >= a + b)
False
>>>
```

When there is an arithmetic expression on the right-hand side of a logical operator, the right-hand side is evaluated first and then the logical operator is applied. The precedence of a relational operator is as shown in table 2.4.

In the table for the precedence of operations, the relational operators have lower precedence than arithmetic operations.

2.2.4   LOGICAL OPERATORS

Table 2.5 is a table for the Python logical operators and as it can be seen, there are three operators in this set. They are *"and"*, *"or"* and *"not"*. The meaning of these operators is the same as we use them in the English language (table 2.5).

2.2.4.1   and Operator

An *and operator* has two operands. Let p and q be two operands, we denote *and* operation by *p and q*. The *and* operation results as true when both propositions are true. Table 2.6 shows the *and* operation.

TABLE 2.4
Precedence of Relational Operator

Operator	Meaning
<=,<,>,>=	Relational operator
==, !=	Equality operator

TABLE 2.5
Logical Operators

Operator	Meaning	Usage
and	True if both the operands are true	x and y
or	True if either of the operands is true	x or y
not	complements the operand	not x

TABLE 2.6
and **Operator**

p	q	p and q
True	True	True
True	False	False
False	True	False
False	False	False

Example 2.21 "today is Friday" and "it is raining".

Let the first operand p be "today is Friday" and the second operand q be "it is raining". The truth value of the *and* operation is true if both p and q are true.

2.2.4.2   or Operator

The *or operator* also has two operands and the result will be true if either of the operands is true. We denote this operation by *p or q*. The *or* operation results as true when at least one of the propositions is true. This is called inclusive-or.

Example 2.22 "students who have taken calculus or computer science can take this course". It means those who have taken calculus or computer science or both can take this course.

Let p be: "those who have taken calculus" and q be: "those who have taken computer science". The corresponding truth table is represented as shown in table 2.7

2.2.4.3   not Operator

The *not operator* is true if the operand is false and vice versa. This is a unary operator.

Example 2.23 "not 50 > 5". Let p is "is 50 > 5". The answer to the question is true. Since the question is "not 50 > 5", so the answer would be false. Table 2.8 shows the *not* operation.

If the operator is true, *not* makes it false, and if it is false, *not* makes it true. The logical operators have the lowest precedence in the precedence table (table 2.14).

TABLE 2.7
or **Operator**

p	q	p or q
True	True	True
True	False	True
False	True	True
False	False	False

TABLE 2.8
not **Operator**

P	not p
True	False
False	True

2.2.5 INCREMENT AND DECREMENT OPERATORS

Python also provides increment operators and decrement operators. Incremental operation adds a number to the variable and saves the result in the same variable (i.e., x = x + 1). In Python x = x + 1 is written as x += 1. The decrement operation subtracts a number from the variable and saves the result in the same variable. (i.e., x = x – 4).In Python x = x – 4 is written as x –= 4.

Example 2.24

```
>>> x = 50
>>> x += 1
>>> print(x)
51
>>> x -= 4
>>> print(x)
47
>>>
```

2.2.6 BITWISE OPERATORS

There are six bitwise operators; *bitwise AND, bitwise OR, bitwise XOR, bitwise NOT*(Complement), *bitwise SHIFT TO THE LEFT* and *bitwise SHIFT TO THE RIGHT*. Table 2.9 shows the bitwise operators.

2.2.6.1 Bitwise AND Operator (&)

In the **bitwise AND operator** (denoted as &), the output is 1 if the corresponding bit of each operand is 1, otherwise, it is 0. To find the result in the bitwise AND operator

TABLE 2.9

Bitwise Operators

Operator	Meaning	Usage
&	Bitwise AND	50 & 51
|	Bitwise OR	50 | 51
^	Bitwise XOR	50 ^ 51
~	Bitwise NOT (Complement)	~ 50
<<	Bitwise Left Shift	10 << 2
>>	Bitwise Right Shift	10 >> 2

TABLE 2.10

Bitwise AND Operation

p	q	p & q
1	1	1
1	0	0
0	1	0
0	0	0

we need to remember the logic AND operator, which is described here. Table 2.10 shows the bitwise AND operation.

Example 2.25 When we find 50 & 51. The output is 50.

```
>>> x = 50
>>> y = 51
>>> z = x & y
>>> print(z)
50
>>>
```

The binary equivalent of 50 is 110010, and the binary equivalent of 51 is 110011. Now take the bitwise AND operation

50 =	0	0	1	1	0	0	1	0
51=	0	0	1	1	0	0	1	1
50=	0	0	1	1	0	0	1	0

Observe that the result is 00110010 which is nothing but 50 in decimal.

TABLE 2.11

Bitwise OR Operation

p	q	p | q
1	1	1
1	0	1
0	1	1
0	0	0

2.2.6.2   Bitwise OR Operator (|)

In the *bitwise OR operator* (denoted as |), the output is 0 if the corresponding bit of each operand is 0, otherwise, it is 1. To find the result in the bitwise OR operator we need to remember the logical OR operator, described in this paragraph. Table 2.11 shows the bitwise OR operation.

Example 2.28 When we find 50 | 51. The output is 51.

```
>>> x = 50
>>> y = 51
>>> z = x | y
>>> print(z)
51
>>>
```

The binary equivalent of 50 is 110010, and the binary equivalent of 51 is 110011. Now take the bitwise OR operation

Observe that the result is 00110011 which is nothing but 51 in decimal.

2.2.6.3   Bitwise XOR Operator (^)

In the *bitwise XOR operator* (denoted as ^), each bit of the output is 1 if the corresponding bit of each operand is different, otherwise, it is 0. To find the result in the bitwise XOR operator we need to remember the exclusive or operator, described in this paragraph. Table 2.12 shows the bitwise XOR operation.

Example 2.29 When we find 50 ^ 51. The output is 1.

```
>>> x = 50
>>> y = 51
```

TABLE 2.12

Bitwise XOR Operation

p	q	p ^ q
1	1	0
1	0	1
0	1	1
0	0	0

```
>>> z = x ^ y
>>> print(z)
1
>>>
```

The binary number of 50 is 110010, and the binary number of 51 is 110011. Now take the bitwise OR operation

50 =	0	0	1	1	0	0	1	0
51 =	0	0	1	1	0	0	1	1
1 =	0	0	0	0	0	0	0	1

Observe that the result is 00000001 which is nothing but 1 in decimal.

2.2.6.4 Bitwise Complement (~)

In the *bitwise complement* (denoted as ~), it returns by *flipping* (converts all 0s to 1s and all 1s to 0s) each bit of the given number. It is basically the same as a negative of the next number to the given number. Negative numbers are stored in 2s complement form.

Example 2.30 When we find ~50. The output is −51.

```
>>> x = 50
>>> z = ~x
>>> print(z)
-51
>>>
```

50 =	0	0	1	1	0	0	1	0
~50=	1	1	0	0	1	1	0	1
-51=	1	1	0	0	1	1	0	1

The binary number of 50 is 00110010. The binary number of 51 is 00110011

1s complement of 51 (flip 1s to 0s and 0s to 1s in 51) = 11001100
2s complement 51 is adding 1 to 1s complement of 51 = 11001100 +1 = 11001101

2.2.6.5 Bitwise Left Shift (<<)

In the **bitwise left shift** (denoted as <<), it takes two operands and returns bits of the first operand shifted left by the second operand places. After left shifting, fill the blank places on the right-hand side with zeros. If x and y are two operands then it is the same as multiplying x by 2 ** y.

Example 2.31 50 << 2 means the binary of 50 is shifted left by 2 places. After shifting left fill the blank spaces with zeros. It is same as 50 * (2^2) = 200.

```
>>> x = 50
>>> y = 2
>>> z = x << y
>>> print(z)
200
>>>
```

The binary number of 50 is 00110010. If 00110010 is shifted by 2 places to the left the result is 11001000, which is nothing but 200 in decimal.

$(50)_2 =$	0	0	1	1	0	0	1	0
Left shift 2 places	1	1	0	0	1	0		
Fill the right-hand blank spaces with 0s	1	1	0	0	1	0	0	0

Example 2.32 5 << 1 means the binary of 5 is shifted left by 1 place. It is the same as 5 * (2^1) = 10.

```
>>> x = 5
>>> y = 1
>>> z = x << y
>>> print(z)
10
>>>
```

The binary of 5 is 0101, with left shift 1 place we get 1010, which is 10 in decimal.

2.2.6.6 Bitwise Right Shift (>>)

In the *bitwise right shift* (denoted as >>), it takes two operands and returns bits of the first operand shifted right by the second operand places. After right shifting, fill the blank places on the left-hand side with zeros. If x and y are two operands then it is the same as dividing x by 2 ** y.

Example 2.33 50 >> 2 the binary of 50 is shifted right
by 2 places. It is the same as 50 / (2^2) = 12.

```
>>> x = 50
>>> y = 2
>>> z = x >> 2
>>> print(z)
12
>>>
```

The binary number of 50 is 00110010. If 00110010 is shifted by 2 places to the
right the result is 00001100, which is nothing but 12 in decimal.

$(50)_2 =$	0	0	1	1	0	0	1	0
Right shift 2 places			0	0	1	1	0	0
Fill the left-hand blank spaces with 0s	0	0	0	0	1	1	0	0

Example 2.34 5 >> 1 means the binary of 5 is shifted
right by 1 place. It is the same as 5 / (2^1) = 2

```
>>> x = 50
>>> y = 1
>>> z = 5 >> 1
>>> print(z)
2
>>>
```

The binary of 5 is 0101, by right shifting 1 place the result is 0010, which is 2 in
decimal.

2.2.7 MEMBERSHIP OPERATORS

There are two membership operators. They are **in** and **not in**. It checks if an element is a member of a sequence such as String, tuple, list or set. The **in** operator
determines if an item is in a sequence, and the *not in* operator is the reverse of the **in**
operator and determines if an item is not in a sequence.

Example 2.35

```
>>> str1 = "Python is Fun"
>>> str2 = "Python"
>>> str3 = "python"
>>> print(str2 in str1)
True
```

```
>>> print(str3 in str1)
False
>>>
```

In this example, we can see that the String *str2* is a substring of *str1* so the in operator returned True. But String *str3* is not a substring of *str1* so the in operator returned False.

Example 2.36

```
>>> x = [1, 2, 3, 4, 5, 6]
>>> y = 5
>>> print(y in x)
True
>>>
```

In this example, we have used the **in** operator to check if 5 is a member of the list. As 5 is a member of the list [1, 2, 3, 4, 5, 6], so it returned **True**.

Example 2.37

```
>>> x = [1, 2, 3, 4, 5, 6]
>>> y = 10
>>> print(y in x)
False
>>>
```

With the same argument, we can see that as 10 is not a member of the list [1, 2, 3, 4, 5, 6] so it returned **False**. The membership operators have very low precedents. Table 2.13 shows the membership operators.

2.2.8 IDENTITY OPERATORS

We check if two variables or two objects are the same, which means if they are pointing to the same location in the memory. Those operators are **is** and **is not**.

TABLE 2.13
Membership Operators

Operator	Meaning	Usage
in	Check to see if the value is in the sequence	5 in [2,5,3,7]
not in	Check to see if the value is not in the sequence	5 not in [2,5,3,7]

Example 2.38

```
>>> x = 5
>>> print(type(x) is int)
True
>>> print(type(x) is not int)
False
>>>
```

In this example, we have written x = 5, which is an Integer datatype. When we checked if *type(x)* is *int*, it returned True. Both the operands point to the same location. Let us see the difference between the == (relational) and *is* (identity) operators.

Example 2.39 The == operator checks for equality in values whereas the "is" operator checks for location

```
>>> x = [1, 'hi' , 2]
>>> y = x
>>> print(x)
[1, 'hi', 2]
>>> print(y)
[1, 'hi', 2]
>>>
```

Observe that the value of x and y are the same.

Example 2.40

```
>>> print(x == y)
True
>>> print(x is y)
True
>>>
```

When we perform *x == y* observe that they have the same values. While comparing the variable with the *is* operator, both variables point to the same list that's why *x is y* returns True.

Example 2.41

```
>>> x = [1, 'hi' , 2]
>>> y = [1, 'hi' , 2]
>>> print(x == y)
True
>>> print(x is y)
False
>>>
```

TABLE 2.14

Precedence Order of the Operators

Operator	Meaning
()	Parenthesis
**	Exponentiation
~,+, −	Complement, unary plus and unary minus
*,/, %, //	Multiply, divide, modulo and whole (floor) division
+, −	Addition and Subtraction
<<, >>	Right and left bitwise shift
&	Bitwise and
^,		Bitwise exclusive or and bitwise or
<=, < ,> ,>=	Relational operators
<> ,== ,!=	Equality operators
= ,%= ,/= ,//= ,-= ,+= ,*=, **=	Assignment operators
is, is not	Identity operators
in, not in	Membership operators

Now we can observe that x is y returns False as x and y are not pointing to the same memory location.

2.2.9 PRECEDENCE OF ALL OPERATORS

The precedence of all the operators is followed as per table 2.14.

2.3 TYPE CASTING

In Python, there is no type declaration. So, the interpreter determines the type of the variable based on the value assigned to it. If we assign x = 5, the interpreter takes x as an Integer variable. Similarly, if we assign x = '5', the interpreter takes x as a String variable. Observe that based on the value assigned, the datatype is determined.

Example 2.42

```
>>> x = 5
>>> print("The type of x is ", type(x))
The type of x is <class 'int'>
>>> x = 5.5
>>> print("The type of x is ", type(x))
The type of x is <class 'Float'>
>>> x = '5'
>>> print("The type of x is ", type(x))
The type of x is <class 'str'>
>>>
```

The type casting done by the interpreter is known as *implicit* type casting. Python interpreters sometimes cannot do type casting on their own. At that time the interpreter throws an error.

Example 2.43

```
>>> x = 5
>>> y = '5'
>>> print("The sum of x and y is ",x + y)
Traceback (most recent call last):
  File "<pyshell#60>", line 1, in <module>
    print("The sum of x and y is ",x + y)
TypeError: unsupported operand type(s) for +: 'int' and 'str'
>>>
```

Observe the last line of the output. It is a *TypeError*. It means we cannot perform the addition operation when one operand is an Integer and the other is a String. In this case, the interpreter doesn't understand whether to convert Integer 5 to String '5' and result in 55 or to convert String '5' to Integer 5 and result in 10. In such a situation *explicit* type casting is required. It means the user needs to do the type casting as desired.

Example 2.44

```
>>> x = 5
>>> y = '5'
>>> print("The sum of x and y is ",str(x) + y)
The sum of x and y is 55
>>> print("The sum of x and y is ",x + int(y))
The sum of x and y is 10
>>>
```

Observe the first print statement *str(x)* converts Integer-valued *x* to String. The output is 55 as a result of String concatenation. In the second print statement, int(y), converts String-valued *y* to an Integer. The output is 10 as a result of the addition of Integers. Here the *explicit* type casting is done. There are some common *explicit* types of casting permitted in Python. Let us try to understand them with examples.

Example 2.45

```
>>> x = 5.5
>>> y = '5'
>>> z = 'a'
>>> print("int(", x,") is ", int(x))
int( 5.5 ) is 5
```

```
>>> print("int(", y,") is ", int(y))
int( 5 ) is 5
>>> print("int(", z,") is ", int(z))
Traceback (most recent call last):
  File "<pyshell#70>", line 1, in <module>
    print("int(", z,") is ", int(z))
ValueError: invalid literal for int() with base 10: 'a'
>>>
```

Observe that int(float) gives an Integer, int(numeric String) gives the number and int(alphabet) gives Error. If the String is a valid numeric String, with a respective base, this can be converted to the respective decimal number on type casting Integer.

Example 2.46

```
>>> print("33 in Hexadecimal is ", int('33',16), " in
Decimal")
33 in Hexadecimal is 51 in Decimal
>>> print("33 in Octal is ", int('33',8), " in Decimal")
33 in Octal is 27 in Decimal
>>> print("33 in Decimal is ", int('33',10), " in Decimal")
33 in Decimal is 33 in Decimal
>>> print("1010 in Binary is ", int('1010',2), " in Decimal")
1010 in Binary is 10 in Decimal
>>>
```

Observe that 33 in hexadecimal is converted to 51 in decimal by *int('33',16)*. Here an int type casting is done whereas Float(x) type casting converts a given numeric String to its corresponding floating-point number.

Example 2.47

```
>>> x = 5
>>> y = 5.5
>>> z = 'a'
>>> print("Float(", x, ") is ", Float(x))
Float( 5 ) is 5.0
>>> print("Float(", y, ") is ", Float(y))
Float( 5.5 ) is 5.5
>>> print("Float(", z, ") is ", Float(z))
Traceback (most recent call last):
  File "<pyshell#80>", line 1, in <module>
    print("Float(", z, ") is ", Float(z))
ValueError: could not convert String to Float: 'a'
>>>
```

Observe that *float(int)* gives a Float, *float(numeric string)* gives the Float and *float(alphabet)* gives Error.

Example 2.48

```
>>> a = 5
>>> b = 5j
>>> c = '5'
>>> d = '5+6j'
>>> print("complex (", a, ") is ", complex(a))
complex ( 5 ) is (5+0j)
>>> print("complex (", b, ") is ", complex(b))
complex ( 5j ) is 5j
>>> print("complex (", c, ") is ", complex(c))
complex ( 5 ) is (5+0j)
>>> print("complex (", d, ") is ", complex(d))
complex ( 5+6j ) is (5+6j)
>>>
```

Observe that *complex(int)* gives a real part of a complex number with an imaginary part as 0, *complex(numeric string)* gives a complex number and *complex(alphabet)* gives Error. There are a few more types of casting available in Python, which we will discuss in due course.

MULTIPLE CHOICE QUESTIONS

1. The *ox* prefix means that the number after it is denoted as
   a. A decimal
   b. A hexadecimal
   c. An octal
   d. A binary

2. The // operator
   a. Performs Integer division
   b. Does not exist
   c. Perform regular division
   d. Performs modular division

3. The result of 123 +0.0 is
   a. 123
   b. 123.0
   c. Cannot be evaluated
   d. None of the above

4. The right-side binding means the expression will be evaluated
   a. From left to right
   b. From right to left
   c. In random order
   d. None of the above

5. The value of the expression 1 ** 2 ** 3 will be evaluated to
   a. 1
   b. 2
   c. 3
   d. 9

6. A keyword is a word that
   a. Is the most important word in the whole program
   b. Cannot be used as a variable name
   c. Is key to puzzles
   d. Is any word in the program

7. The output of print(float('100.0')) is
   a. 100.0
   b. 100
   c. Error
   d. None of the above

8. Which one of the following is true?
   a. Addition precedes multiplication
   b. Multiplication precedes addition
   c. Any random order
   d. Neither statement can be evaluated

9. The output of print(True and (3<=5)) is
   a. False
   b. True
   c. FALSE
   d. TRUE

10. Find the output of the following code
    x = 'Python Course'
    print('course' in x)

    a. False
    b. True
    c. FALSE
    d. TRUE

11. Find the output of the following code
    x = 0
    print(not x)

    a. True
    b. False
    c. true
    d. false

12. Find the output of the following code
    X = 6
    print(4 * 2 – 2 ! = x)

    a. True
    b. False
    c. true
    d. false

13. Find the output of the following code
    x = 4
    x == 3 + 2
    print(x)

    a. 5
    b. 4
    c. True
    d. False

14. Find the output of the following code
    x = 4
    print(x == 3 + 2)

    a. 5
    b. 4
    c. True
    d. False

15. Find the output of the following code
    print(15 // 2)

    a. 5
    b. 6
    c. 7
    d. 7.5

16. Find the output of the following code
    print(15 << 2)

    a. 30
    b. 45
    c. 60
    d. 75

17. Which of the following options refer to the code?
    print(15 << 2)

    a. 15 * 2
    b. 15 * (2 + 2)
    c. 15 * (2 ** 2)
    d. 15* (2 + 3)

18. Which of the following options refer to the code?
    print(15 >> 2)

    a. 1
    b. 2
    c. 3
    d. 4

19. Which of the following options refer to the code?
    print(15 >> 2)

    a. 15 / 2
    b. 15 / (2+2)
    c. 15 / (2**2)
    d. 15/(2+3)

20. The value of the expression 16 % 7 % 4 will be evaluated to
    a. 1
    b. 2
    c. 3
    d. 0

DESCRIPTIVE QUESTIONS

1. Describe the comparison operators.
2. What are identity operators and membership operators?
3. What is floor division? Explain with an example.
4. Describe bitwise operators with examples.
5. Write the precedence rules of arithmetic operators.
6. Describe logical operators and their precedence.
7. Describe modulo operations and write about how the modulo operator in python is different from other programming languages.

PROGRAMMING QUESTIONS

1. Write a Python program that calculates compound Interest.
2. Write a Python program that swaps two numbers using a third variable.
3. Write a program that asks the user to enter a positive Integer n (in seconds). Your program should convert the number of seconds into days, hours, minutes and seconds.
4. Write a Python program that takes input as a binary String and converts it to a decimal number (use print).

ANSWER TO MULTIPLE CHOICE QUESTIONS

1.	B	2.	A	3.	B	4.	B	5.	A
6.	B	7.	A	8.	B	9.	B	10.	A
11.	A	12.	B	13.	B	14.	D	15.	C
16.	C	17.	C	18.	C	19.	C	20.	B

3 Control Structures

LEARNING OBJECTIVES

After studying this chapter, reader will be able to:

- Identify and use various conditional statements
- Understand the importance of indentation
- Identify and use various looping-structures
- Identify when to use for-loop and when to use while loop
- Use break, continue and pass
- Use the range function
- Use nested loop

video on **Control Structures**

3.1 INTRODUCTION

Until now, whichever programs we have written so far in this book, the statements were executed sequentially. In everyday life, one has to make some decisions based on some conditions or one has to do certain work repeatedly for different objects. Similarly, in programming one also needs to change the sequence of execution of the statements. In many other situations, one needs to execute a sequence of statements repeatedly. The flow control refers to the order in which the statements in a code are executed.

The control flow of statements is broadly classified as branching and looping structure. Branching statements are also known as conditional statements. They alter the sequence of program statements. The branching statements are:

1. if-else statement
2. if-elif-else statement

The looping structure or iterating statements are the sections of code that gets executed repeatedly until some termination condition is satisfied. The looping statements are:

1. for-loop
2. while-loop

If a condition is satisfied during branching or looping, a particular block of instructions will be carried out. These blocks in Python are expressed by indentation. The rule of expressing a block of code by indentation is known as *off-side notation* for

DOI: 10.1201/9781003219125-4

coding. The colon (:) symbol is used as a delimiter followed by a tabbed block of code.

> ***Branching/Looping statement :***
> ***Block for Branching/Looping statement***
> ***Continuation after the Branching/Looping***

3.2   CONDITIONAL STATEMENTS

Let us try to understand the conditional statements with the help of examples. Let us take a statement "Srusti completes reading this book before going to sleep". By this statement, it means "If Srusti completes reading this book, she will go to sleep" and this is a conditional statement.

3.2.1   CONDITIONAL IF-ELSE

In computer programming, conditional statements are used to perform different computations or actions depending on whether the condition is True or False.

This is the syntax for an *if* statement in Python:

> ***if Conditional_Expression :***
> ***Statement(s)1***
> ***else:***
> ***Statement(s)2***

Here "*if*" and "*else*" are the keywords. *Conditional_Expression* should always evaluate to either True or False followed by a colon symbol (:). If the Conditional_ Expression evaluates to True then Statement(s)1 is to be executed. If the Conditional_ Expression is evaluated to be False then Statement(s)2 followed by *else* is to be executed. The Conditional_Expression may be kept inside a pair of brackets ( ). However, the brackets are optional. For a proper understanding of complicated Conditional_Expression, it is advisable to keep proper parenthesis. Figure 3.1 shows the flow diagram of conditional if-else expression.

In the if-else statement notice that:

- The *if-else* statement starts with the keyword *if.*
- The *Conditional_Expression* returns either True or False.
- A *colon* (:) is always required. The next line after the colon starts the body of the if statement, which consists of one or more statements. And all these statements must be tabbed. Any statement which is not tabbed and follows the body of the if-else is not part of it.
- When the *Conditional_Expression* returns True, the statements in the body of the **if** are executed.
- When the *Conditional_Expression* returns False, the statements in the body of the **else** are executed.
- The *else* is optional in the if-else statement.

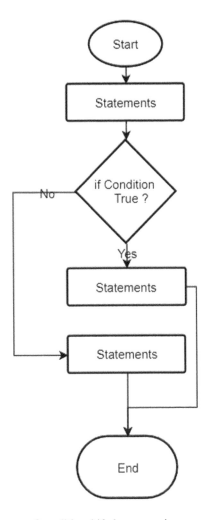

FIGURE 3.1   Flow diagram of conditional if-else expression

Example 3.1

```
if 8 > 5:
  print (" 8 > 5 is True")
else:
  print (" 8 > 5 is False")
```

In this case, the condition "8 > 5" is TRUE, Python executes the line which is coming after the *if*. That is *8 > 5 is* True is printed.

Output

```
8 > 5 is True
```

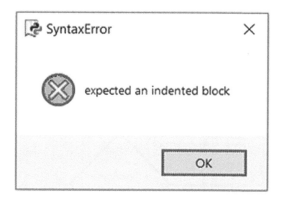

FIGURE 3.2   Error message due to improper indentation

If the indentation is not correct, the *expected an indented block* error messages appear as shown in figure 3.2.

Example 3.2

```
if 8 > 5:
  print (" 8 > 5 is True")
else:
  print (" 8 > 5 is False")
```

For a better understanding of the importance of the tabbed statement (indented statements), let us take some more examples.

Example 3.3

```
my_list = ["Hi","How","are","You"]
my_input = input ("Enter your String ")
if my_input in my_list:
  print ('yes')
else:
  print ('no')
  print ('Better Luck Next Time')
```

When the input is given is "Hi", this is in my_list. So it executed the if part and printed 'yes'

Output 1

```
Enter your String Hi
yes
```

When the condition *my_input in my_list* is False, the print statements after else got executed. Here the input given is 'hi', this is not in *my_list*. So, it executed the statements, *print('no')* and *print('Better Luck Next Time')*, which is inside the else part.

Output 2

```
Enter your String hi
no
Better Luck Next Time
```

3.2.2 CONDITIONAL IF-ELIF-ELSE

When multiple conditions are to be checked, we use if-elif-else statements. Notice that "else if" is written as *elif* in Python.

This is the syntax for "if-elif-else" statement in Python is:

if Conditional_Expression1 :
    Statement(s)1

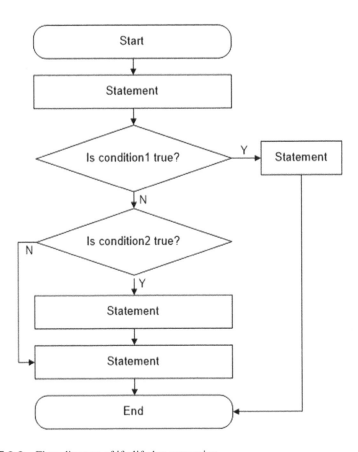

FIGURE 3.3   Flow diagram of if-elif-else expression

elif Conditional_Expression2 :
    Statement(s)2
else:
    Statement(s)3

The syntax of the *if-elif-else* statement looks the same as an *if-else* statement. Figure 3.3 shows the flow diagram of an *if-elif-else* expression.

- It starts with a keyword, *if* and then it is followed by *Conditional_ Expression1* and a colon (:).
- If the *Conditional_Expression1* evaluates to True, Statement(s)1 are executed. Make sure that Statement(s)1 have tabs before them.
- If the *Conditional_Expression1* evaluates to False, the keyword *elif* is encountered.
- The *elif* statement is followed by *Conditional_Expression2* and a colon (:). When the Conditional_Expression2 evaluates to True, Statement(s)2 are executed. Make sure that Statement(s)2 have tabs before them.
- If the Conditional_Expression2 also evaluates to False, the keyword "else" is encountered.
- This is the last block of statements. Notice that after the "else", there is no expression. It simply is followed by a colon.
- It means when the *Conditional_Expression1* and *Conditonal_Expression2* evaluate to False, the *else* block gets executed.

Example 3.4

```
my_list1 = ["Hi", "How", "are", "You"]
my_list2 = ["I", "am", "Fine"]
my_input = input("Enter your String ")
if (my_input in my_list1):
  print(my_input, "is in LIst1")
elif (my_input in my_list2):
  print(my_input, 'is in List2')
else:
  print(my_input, 'is NOT available in any List')
```

In this sample program, we have defined two lists as my_list1 = ["Hi", "How", "are", "You"] and my_list2 = ["I", "am", "Fine"]. If the first conditional statement i.e., if(my_input in my_list1) is True, it will execute print(my_input, 'is in List1'). If the first conditional statement is FALSE it will check the *elif* part i.e., elif(my_input in my_list2). If the second condition is True, it will execute print(my_input, 'is in List2'). At last, when both the conditions are False, it will execute the last statement print(my_input, 'is NOT available in any List'). This can be verified from the output.

Output 1

```
Enter your String Hi
Hi is in LIst1
```

Output 2

```
Enter your String am
am is in List2
```

Output 3

```
Enter your String 3
3 is NOT available in any List
```

Here there are two things to notice. The first one is, after the *if* statement, there's a conditional expression and a colon. After *elif*, there's a conditional expression and a colon. And after *else*, there is no expression but a colon. The second one is, that each block of statements must be tabbed.

Example 3.5 Write a program that accepts an Integer as a percentage of marks. It will determine whether the percentage of marks is 1st division, 2nd division or 3rd division.

```
my_input = int(input("Enter Percentage "))
if (my_input < 50)  :
  print(my_input, 'is Third Division')
elif(my_input >= 50 and my_input <60 ):
  print(my_input,'is Second Division')
else:
  print(my_input,'is First Division')
```

Output 1

```
Enter Percentage 45
45 is Third Division
```

Output 2

```
Enter Percentage 55
55 is Second Division
```

Output 3

```
Enter Percentage 65
65 is First Division
```

Example 3.6 Write a program that accepts the lengths of three sides of a triangle as inputs. The program output should indicate whether or not the triangle is a right triangle

```
A = int(input("Enter side 1 of the Triangle "))
B = int(input("Enter side 2 of the Triangle "))
C = int(input("Enter side 3 of the Triangle "))
```

```
if ((a * a) == (b * b) + (c * c)):
  print("Triangle with sides ",a,", ",b,", ",c," is a right
  angle Triangle")
elif ((b * b) == (c * c) + (a * a)):
  print("Triangle with sides ",a,", ",b,", ",c," is a right
  angle Triangle")
elif ((c * c) == (a * a) + (b * b)):
  print("Triangle with sides ",a,", ",b,", ",c," is a right
  angle Triangle")
else :
  print("NOT a Right angle Triangle")
```

Output 1

```
Enter side 1 of the Triangle 3
Enter side 2 of the Triangle 4
Enter side 3 of the Triangle 5
A triangle with sides 3, 4 or 5 is a right-angle Triangle
```

Output 2

```
Enter side 1 of the Triangle 2
Enter side 2 of the Triangle 3
Enter side 3 of the Triangle 4
NOT a right-angle Triangle
```

Example 3.7 Write a Python program to convert temperatures to and from Celsius, Fahrenheit.

```
temp = input("Input the temperature you like to convert ?
(e.g. 45F, 102C) ")
degree = int(temp[:-1])
i_conversion = temp[-1]
if ((i_conversion == 'c' )|(i_conversion == 'C' )):
    result = int(round((9 * degree)/5 + 32))
    o_conversion = "Fahrenheit"
elif ((i_conversion == 'f' )|(i_conversion == 'F' )):
    result = int(round(( degree -32) * 5 / 9))
    o_conversion = "Celsius"
else :
  print("Input is not in proper format")
  quit()
print("The temperature in ",o_conversion, "is ", result,
"Degrees")
o_conversion = "Fahrenheit"
```

Output 1

```
Input the temperature you like to convert ? (e.g., 45F, 102C) 45f
The temperature in Celsius is 7 Degrees
```

Output 2

```
Input the temperature you like to convert? (e.g., 45F, 102C) 7C
The temperature in Fahrenheit is 45 Degrees
```

Output 3

```
Input the temperature you like to convert? (e.g., 45F, 102C) 54F
The temperature in Celsius is 12 Degrees
```

Output 4

```
Input the temperature you like to convert? (e.g., 45F, 102C) 12c
The temperature in Fahrenheit is 54 Degrees
```

3.2.3   NESTED IF-ELIF-ELSE STATEMENTS

It is possible to have an if-elif-else structure inside another if-elif-else structure. This is called *nested if* statement. Any number of if-elif-else statements can be inside other if-elif-else statements. The level of nesting is identified by observing the indentation. Any change in indentation may raise an error or may change the meaning of the statement. So it is advisable to avoid using the nested if structure.

Example 3.8

```
age = int(input("Enter your age:"))
if (age > 12):
  if(age < 20):
    print("You are in teenage")
  else:
    print("You have crossed teenage")
else:
  print("You will reach teenage soon")
```

Observe the statement *if(age < 20)* , which is inside *if(age > 12)* statement. It means the inner *if* statement is executed once the outer returns True.

Output 1

```
Enter your age:10
You will reach teenage soon
```

Output 2

```
Enter your age:15
You are in teenage
```

Output 3

```
Enter your age: 23
You have crossed teenage
```

In the above example when the outer and inner if statements are True, then only *You are in teenage* will be printed. In order to avoid the nested if structure, we can use logical *and* operations. The same example can be written as:

Example 3.9

```
age = int(input("Enter your age:"))
if (age > 12 and age < 20):
  print("You are in teenage")
elif (age < 12):
  print("You will reach teenage soon")
else:
  print("You have crossed teenage")
```

Output 1

```
Enter your age:10
You will reach teenage soon
```

Output 2

```
Enter your age:15
You are in teenage
```

Output 3

```
Enter your age: 23
You have crossed teenage
```

3.2.4 TERNARY OPERATOR

The ternary operator is a one-line *if-else* statement. Instead of using multi-line *if-else* statements, a ternary operator is used.

The syntax for a simple "ternary operator" statement is:

Expression1 if Conditional_Expression else Expression2

Here "if " and "else" are the keyword. "Conditional_Expression" should always evaluate to either True or False. If the Conditional_Expression evaluates to True then *Expression1* is to be executed and if the Conditional_Expression is evaluated to be False then *Expression2* is to be executed. The "Conditional_Expression" may be kept inside a pair of braces ( ). However, the braces are optional. For a better understanding of complicated Conditional_Expression, it is advisable to keep proper parenthesis.

Example 3.10

```
a = int(input("Enter First Number "))
b = int(input("Enter First Number "))
```

```
print ("The bigger number is:", a) if a > b else print("The
bigger number is:",b)
```

It is also possible to assign a variable to the *expression*.

Example 3.11

```
a = int(input("Enter First Number "))
b = int(input("Enter First Number "))
Max = a if a > b else b
print ("The bigger number is:", Max)
```

Tuples (refer to Chapter 5) can be used to implement a ternary operator.
The general syntax is:

(Variable2/Value 2 on False, Variable1/Value 1 on True) [Conditional_Expression]

The conditional statement is written inside the square bracket and the result of the conditional statement is written within a pair of parenthesis just before the square bracket. The conditional expression returns either True or False, so there should be exactly two values inside the parenthesis. Where the first variable/value is assigned if the condition is evaluated to False and the second variable/value is assigned if the condition is evaluated to True.

Example 3.12

```
a = int(input("Enter First Number "))
b = int(input("Enter First Number "))
Max = (b, a)[a > b]
print ("The bigger number is:", Max )
```

Dictionary (refer to Chapter 6) can also be used to implement a ternary operator.
The general syntax is:

{False: Variable2/Value 2 , True :Variable1/Value1 } [Conditional_Expression]

The conditional statement is written inside the square bracket and the result of the conditional statement is written within a pair of curly brackets just before the square bracket. The conditional expression returns either True or False, so there should be exactly two values inside the curly braces. The first pair is *False: Variable2/Value 2* indicates if the conditional Expression is evaluated to be False, Variable2/Value2 is returned. The second pair is *True: Variable1/Value1* indicates if the conditional Expression is evaluated to be True, Variable1/Value1 is returned.

Example 3.13

```
a = int(input("Enter First Number "))
b = int(input("Enter First Number "))
```

```
Max = {False:b, True:a}[a > b]
print ("The bigger number is:", Max )
```

Output

```
Enter First Number 3543
Enter First Number 84
The bigger number is: 3543
```

Just like the nested if-else statement it is also possible to write the nested ternary operator. The general syntax is:

Expression1 if Conditional_Expression_1 else Expression2 if Conditional_Expression_2 else Expression3

The *Conditional_Expression_1* is first evaluated. If the condition is evaluated to be True, it returns *Expression1* otherwise it evaluates *Conditional_Expression_2*. If this expression is evaluated to be True, it returns *Expression2* otherwise it returns *Expression3*.

Example 3.14

```
a = int(input("Enter First Number "))
b = int(input("Enter Second Number "))
Max = a if a > b else b if b > a else "Both are Equal"
print ("The bigger number is:", Max )
```

First, the condition a > b is evaluated, if the condition is True, it returns "a" otherwise it checks the condition b > a. And if this condition is evaluated to True, it returns "b" otherwise returns *Both are Equal*.

Output 1

```
Enter First Number 2
Enter Second Number 3
The bigger number is: 3
```

Output 2

```
Enter First Number 3
Enter Second Number 2
The bigger number is: 3
```

Output 3

```
Enter First Number 3
Enter Second Number 3
The bigger number is: Both are Equal
```

3.3 LOOPS

Looping refers to iterated (repeated) execution of statements. There are two different types of looping structures in Python.

video on **Looping Structure**

1. while-loop
2. for-loop

3.3.1 WHILE-LOOP

The while-loop repeatedly executes a set of statements as long as the conditional_ Expression is true. This looping structure is called conditional as we don't know in advance the boundary for terminating the loop in advance. Figure 3.4 shows the flow diagram of the while expression.

The syntax for while-loop is:

while conditional_Expression:
    Statement(s)1
else:
    Statement(s)2

In the while-loop notice that:

- The while-loop starts with the keyword *while*.
- The *conditional_Expression* evaluates either True or False. It is not neces-sary to have the conditional_Expression inside a pair of parenthesis.
- A *colon* (:) is always required. The next line after the colon starts the body of the while-loop, which consists of one or more statements. And all these statements must be tabbed. Any statement which is not tabbed and follows the body of the while-loop is not part of the while-loop.
- When the *conditional_Expression* evaluates True, the statement(s)1 in the body of the while-loop are executed.
- Also, be very careful, if the conditional statement in the while-loop always evaluates to be true (never becomes false), then the loop never ends. The loop which never comes to an end is called an infinite loop.
- The **else** in the while-loop is optional. When the value of the control vari-able is not in the *Collection_of_items*, it executes the statement(s)2 inside the else block.

The while-loop is used when:

- We need to repeatedly execute some statements until some condition is met.
- We don't know in advance when this condition will be met.

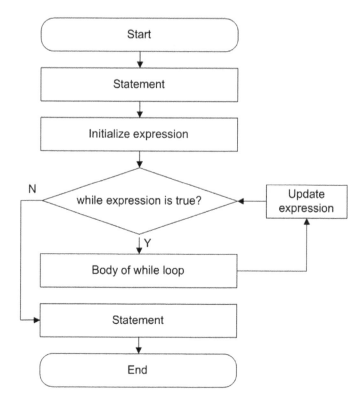

FIGURE 3.4   Flow diagram of while expression

Notice similarities with the *if* statement, we used the keyword *while* instead of the keyword *if*. The semantic difference is more important: in *if* statements, when the condition is met, it performs all the statements inside the body only once; whereas *while* repeats the execution of all the statements inside the body as long as the condition evaluates to True.

Example   3.15   Write   a   program   that   will   print   a   simple statement 5 times.

```
i = 0
while i < 5:
  print(i, "I \'m Learning Python")
  i = i + 1
else:
```

```
print ("inside the while-else")
print (i,"I \'m Enjoying it.")
```

Output

```
0 I 'm Learning Python
1 I 'm Learning Python
2 I 'm Learning Python
3 I 'm Learning Python
4 I 'm Learning Python
inside the while-else
5 I 'm Enjoying it.
```

In this example, notice that i is the control variable with $i < 5$ followed by a colon (:). The next line after the colon is a print statement *print(I, "I\'m Learning Python")*. The next statement is $i = i + 1$. We can observe that in the while-loop the iteration starts at $i = 0$ and ends at $i = 4$. When the value of i is 5, the condition is False, now the *else* part will be executed.

The code inside the *else* block gets executed once the code inside *while* block finish executing normally (without a break statement). Observe that the value of "i" is 5 in the else block. It means the control variable is incremented at the end of the while-loop.

Example 3.16 Write a program that prints the sum of all the numbers from 1 to a number provided by the user, that is divisible by 5.

```
my_num = int (input ("Enter a Number: "))
i = 1
sum = 0
while (i <= my_num) :
  if i % 5 == 0:
    sum = sum + i
  i = i + 1
print ("The sum is ",sum)
```

Observe that the conditional_Expression $i <= my\_num$ is kept inside a pair of parenthesis. It is there to improve the readability as such in Python it is not mandatory to have a pair of parenthesis.

Output

```
Enter a Number: 300
The sum is 9150
```

Observe the steps in this while loop. The first statement reads a number from the keyboard provided by the user. The second statement is to initialize from where to start. As the starting number is 1. We need to take a variable assigned to 1 i.e. *i=1*. One more variable we need that will keep the sum of numbers assigned to 0 i.e., *sum=0*.

The next statement is *while(i <= my_num)*. When "the value of *i* is less than the number provided by the user" the conditional expression evaluates to True and then checks if that number is divisible by 5. If the value of *i* is divisible by 5, add *i* to the previous value of the sum to get a new value of the sum. In the next step, increment *i* by 1. In the last step print the sum. Observe that this print statement is outside the while loop.

Example 3.17 Write a program using a while loop which asks the user to type a positive Integer n and then prints the factorial of n. A factorial is defined as the product of all the numbers from 1 to n (1 and n inclusive).

```
my_num = int(input("Enter a Number: "))
fact = 1
i = 1
while (i <= my_num):
  fact = fact * i
  i = i + 1
print("The factorial of ", my_num," is ", fact)
```

Output

```
Enter a Number: 30
The factorial of 30 is 265252859812191058636308480000000
```

3.3.2 FOR-LOOP

The for-loop is yet another method to iterate (repeated) execution of statements. The "for" loop is used in two different situations; the first case is when we want to go over every item in a sequence and the second case is when we want to repeat some action for a given number of times. This looping structure is known as a definite iteration as there is a definite boundary for the termination of the loop. Figure 3.5 shows the flow diagram of a *for* loop.

The syntax of a *for* loop is:

> *for Variable_Name in Collection_of_Items:*
>   *Statement(s)1*
> *else:*
>   *Statement(s)2*

In the for loop notice that:

- The for-loop starts with the keyword *for*.
- Any Variable_Name after the *for* keyword is the *control variable* (or iterating variable) of the loop; it counts the loop's turns and does it automatically.
- The **in** keyword allows taking the element described in the *Collection_of_ items* and the values being assigned to the control variable.

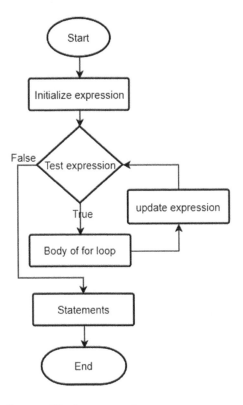

FIGURE 3.5 Flow diagram of for-loop expression

- The *Collection_of_items* can be taken from the *range()* function (this is a very special function) responsible for generating all the desired values of the control variable. This function accepts only integers as its arguments and generates sequences of integers.
- The *Collection_of_items* can also contain a *Start*, which is the first value in the range; if omitted, the default value is 0. An *End*, which is the last value in the range; the end value may not be omitted. And *Increment/Decrement*, if this parameter is omitted, it defaults increment to 1.
- A *colon* (:) is always required. The next line after the colon starts the body of the for-loop, which consists of one or more statement(s)1. And all these statements must be tabbed. Any statement which is not tabbed and follows the body of the for-loop is not part of the for-loop.
- The **else** in the for-loop is optional. When the value of the control variable is not in the *Collection_of_items*, it executes the statement(s)2 inside the else block.

The for-loop is used when:

- We know the maximum number of times the body of the loop is executed.
- We want to iterate over all the elements of a collection such as String, list, tuple, etc.

3.3.2.1 The range() Function

Before doing any example, it is necessary to discuss the *range()* function. The range() function takes three parameters, of which two are optional. The first parameter is *Start*, which is the first value in the range; if omitted, the default value is 0. The second parameter is *End*, which is the last value in the range; the end value may not be omitted. And *Increment/Decrement*, if this parameter is omitted, it defaults the increment to 1.

If only one parameter is given, it refers to the *End* parameter. In this case, the *Start* parameter by default is 0. And the increment by default is 1. It will take all the values from 0 through (End −1).

If only the first two parameters are given, it refers to all the values in the range Start through (End −1) with increment 1. Remember to get some data, the Start must be less than the End, otherwise, the range function will not get executed.

If all the three parameters are given, it refers to all the values in the range Start through (End −1) with increment as the third parameter. If the value in the third parameter is positive, it is called increment. In this case, the Start should be less than the End. If the value in the third parameter is negative, it is called decrement. In this case, the End should be less than the Start.

Examples in table 3.1 show how the range() function can be used to produce a variety of sequences.

Example 3.18 Range function (see table 3.1).

TABLE 3.1

Examples of Range Functions with Output

Range Function	Range Object (Output)
range(10)	0, 1, 2, 3, 4, 5, 6, 7, 8, 9
range(1, 10)	1, 2, 3, 4, 5, 6, 7, 8, 9
range(1, 10, 2)	1, 3, 5, 7, 9
range(10, 0, −1)	10, 9, 8, 7, 6, 5, 4, 3, 2, 1
range(10, 0, −2)	10, 8, 6, 4, 2
range(2, 11, 2)	2, 4, 6, 8, 10
range(−5, 5)	−5, −4, −3, −2, -1, 0, 1, 2, 3, 4
range(1, 2)	1
range(1, 1)	(empty)
range(1, −1)	(empty)
range(1, −1, −1)	1, 0
range(0)	(empty)

Example 3.19 Write a program that will print a simple statement 5 times.

```
for i in range(5):
  print(i, "I \'m Learning Python")
else:
  print("inside the for-else")
  print(i,"I \'m Enjoying it.")
```

Output

```
0 I 'm Learning Python
1 I 'm Learning Python
2 I 'm Learning Python
3 I 'm Learning Python
4 I 'm Learning Python
inside the for-else
4 I 'm Enjoying it.
```

In this example, notice that *i* is the control variable with range(5) followed by a colon (:). The next line after the colon is a print statement that prints "*I 'm Learning Python*" 5 times. We can observe that in the for-loop the iteration starts by default with 0 and ends at 4. It means the range(n) goes from 0 to (n−1). After executing the loop 5 times the value of "i" is 4.

The code inside the *else* block gets executed once the code inside *for* block finish executing normally (without a break statement). Observe that the value of "i" is 4 in the else block also. It means the control variable does not get incremented at the end of the for-loop.

Example 3.20 Write a program that will find the sum of first 9 numbers first

```
sum = 0
for i in range(1, 10):
  sum += i
print("The sum is ", sum)
```

Notice that the range is starting at 1 and ends at 10. The increment is not defined so by default it is 1. This program sums numbers from 1 to 9. Remember that the start value of the range function does need not be 0. We can start with any number that we want.

Output

```
The sum is 45
```

Example 3.21 Write a program that will find the sum of even numbers in the range of 10.

```
sum = 0
for i in range(2, 10, 2):
  sum += i
print("The sum is ", sum)
```

Output

```
The sum is 20
```

Example 3.22 Write a program that will find the sum of odd numbers in the range of 9 with decrement.

```
sum = 0
for i in range(9, 0, -2):
  sum += i
print("The sum is ", sum)
```

Notice that odd numbers are generated starting with 9 through 1. On these data, the sum generated is 25.

Output

```
The sum is 25
```

Example 3.23 Write a program that will access all the elements of a list.

```
my_list = [1, 'Horse', 2.4, 'Tiger', 3, 'Lion']
for i in my_list:
  print(i, "is an item in the list")
```

Output

```
1 is an item in the list
Horse is an item in the list
2.4 is an item in the list
Tiger is an item in the list
3 is an item in the list
Lion is an item in the list
```

In this example, observe that the control variable *i* takes the data from the list one by one till the end of the list. If the item number of the list is to be printed, then we need to take one more variable.

Example 3.24 Write a program that will access all the elements of a list.

```
my_list = [1, 'Horse', 2.4, 'Tiger', 3, 'Lion']
j = 1
```

```
for i in my_list:
  print("Item ",j," in the list is ",i)
  j += 1
print("End of List")
```

Output

```
Item 1 in the list is 1
Item 2 in the list is Horse
Item 3 in the list is 2.4
Item 4 in the list is Tiger
Item 5 in the list is 3
Item 6 in the list is Lion
End of List
```

Notice that a variable j is assigned to 1. This variable is incremented in order to keep track of the item number in the list. The print statement *"End of the List"* is printed after the entire item in the list is printed. And this statement is out of the scope of for loop.

3.3.3 CONTINUE, BREAK AND PASS

Sometimes it is necessary to perform some special tasks like skipping some values in the loop, coming out of the loop, or to do nothing for certain values in the current iteration or loop. In those situations, we can use continue, break and pass statements.

3.3.3.1 Continue Statement

The *continue* statement is used to skip the rest of the code inside the loop for the current iteration only. The loop does not terminate but continues with the next iteration. The continue statement is used as shown in figure 3.6.

The continue statement has the keyword **continue**. Whenever this keyword appears the control simply, without doing anything, goes on executing for the next value of the *Control_Variable*. The usage of continue is the same in both for-loop and while-loop.

Let us discuss an example that uses a continue statement.

```
→for i in range(5):       →while (condition) :
      statement(s)1             statement(s)1
      if (condition):          if (condition):
 └────────continue        └────────continue
      statement(s)2            statement(s)2
```

FIGURE 3.6   Continue statement

Example 3.25 Write a program using a for-loop demonstrating the continue statement

```
for i in range(5):
  if(i == 3):
```

```
     continue
   print(i)
print("Out of Loop")
```

Example 3.26 Write a program using while-loop demonstrating continue Statement

```
i = 0
while i in range(5):
  print(i)
  i += 1
  if(i == 3):
    continue
print("Out of Loop")
```

When the value of *Control_Variable* is *i == 3*, the *continue* statement gets executed. So the value *i = 3* is skipped (not printed). It means all the values from 0 through 4 except 3 are printed. And at last, the statement which is outside the looping structure *print("Out of Loop")* is printed. Generally, the range function is not used in a while loop.

Output

```
0
1
2
3
4
Out of Loop
>>>
```

Example 3.27 Write a program using a for-loop demonstrating continue Statement with String.

```
str = "I AM LEARNING PYTHON"
for i in range(len(str) - 1):
  if (str[i] == 'N'):
    continue
  print(str[i])
print("Out of Loop")
```

Example 3.28 Write a program using a while-loop demonstrating continue Statement with String.

```
str = "I AM LEARNING PYTHON"
i = 0
while (i <= len(str) - 1):
```

```
    print (str[i])
    i += 1
    if (str[i] == 'N'):
      i += 1
      continue
print ("Out of Loop")
```

Notice that the character *N* is missing. Whenever 'N' appeared the continue statement was executed. So the program prints *I AM LEARIG PYTHO*. At last, the control came out of the loop and printed the line which is outside the loop *print("Out of Loop")*.

Output

```
I
A
M
L
E
A
R
I
G
P
Y
T
H
O
Out of Loop
>>>
```

3.3.3.2 Break Statement

The *break* statement terminates the loop containing it. On encountering a break statement inside a loop, the control jumps to the statement immediately after the body of the loop. The *break* statement is used as shown in figure 3.7.

The break statement has a keyword **break**. Whenever this keyword appears the control simply exits the looping structure and executes the next statement immediately after the loop. The usage of break is the same in both for-loop and while-loop.

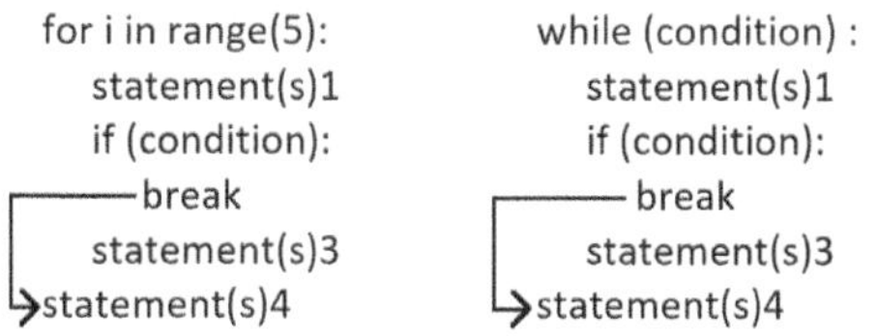

```
        for i in range(5):            while (condition) :
            statement(s)1                 statement(s)1
            if (condition):               if (condition):
           ┌────── break                 ┌────── break
           │   statement(s)3             │   statement(s)3
           └▸statement(s)4               └▸statement(s)4
```

FIGURE 3.7 Break statement

Example 3.29 Write a program using a for-loop demonstrating a break statement.

```
for i in range(5):
  if(i == 3):
    break
  print(i)
  i += 1
print("Out of Loop")
```

Example 3.30 Write a program using a while-loop demonstrating a break statement.

```
i = 0
while i in range(5):
  if (i == 3):
    break
  print(i)
  i += 1
print("Out of Loop")
```

Observe that when the value of *i* is 3, the *break* statement gets executed. On encountering the break statement control comes out of the loop and prints the line which is outside the loop *print("Out of Loop")*.

Output

```
0
1
2
Out of Loop
>>>
```

Example 3.31 Write a program using a for-loop demonstrating a break statement with String.

```
str = "I AM LEARNING PYTHON"
i = 0
for i in range(len(str) - 1):
  if(str[i] == 'N'):
    break
  print(str[i])
```

Example 3.32 Write a program using a while-loop demonstrating a break statement with the String.

```
str = "I AM LEARNING PYTHON"
i = 0
```

```
while i in range(len(str) - 1):
  print(str[i])
  i += 1
  if(str[i] == 'N'):
    break
```

When for the first time 'N' appeared the *break* statement was executed. In other words, characters in the String "*I AM LEARNING PYTHON*", When the value of *Control_Variable* is *str[i]=='N'* encountered, a break gets executed. Now the control comes out of the looping structure. So the output is I AM LEAR.

Output

```
I
A
M
L
E
A
R
>>>
```

Observe that when a break statement is executed it completely gets out of the loop, and executes the next statement which is outside the body of the loop.

3.3.3.3 Pass Statement

The *pass* statement refers to doing nothing (no code). It acts as a placeholder. It means instead of writing nothing, we write pass.

Example 3.33 Write a program using a for-loop demonstrating a pass statement.

```
for i in range(5):
  if(i == 3):
    pass
  else:
    print(i)
print("Out of Loop")
```

Example 3.34 Write a program using a while-loop demonstrating a pass statement.

```
i = 0
while i in range(5):
  if(i == 3):
    pass
  else:
    print(i)
```

```
i += 1
print("Out of Loop")
```

Observe that we have used the *pass* statement when *i* == *3*. When the control encounters the keyword pass, the interpreter knows that there are no instructions to execute.

Output

```
0
1
2
4
Out of Loop
>>>
```

It is possible to get confused with the *pass* and the *continue* statements. The *continue* statement is only used to skip a looping structure but *pass* is used widely when there are no instructions to execute. *Pass* statements can be used in looping structure, functions and class, etc.

3.3.4 NESTED LOOPS

When a loop is inside another loop we say it is a *nested loop*. Any type of loop can be nested under any other loop. This means we can have a *for*-loop inside another *for*-loop, or a *for*-loop inside a *while*-loop or a *while*-loop nested inside another *while*-loop, or we may also have a *while*-loop nested inside a *for*-loop. Figure 3.8 shows a nested loop.

```
for i in range(5):
    for j in range(5) :
        statement(s)1
    statement(s)2
```

FIGURE 3.8   Nested loops

Example 3.35 a for-loop inside another for-loop

The outer for-loop has a control variable *i*. In the body of the outer loop, we have an inner for-loop with a control variable *j*. Notice that the body of the inner for-loop is also tapped like the body of the outer for-loop. For a single value of *i*, the inner-loop executes for all the values of *j*.

Example 3.36 a for-loop inside a while-loop

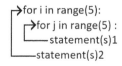

```
i = 0
while(i <= 10):
    for j in range(1,10):
        print(i," x ", j," = ", i * j)
    i += 1
    print("Multiplication table for ",i," completed")
```

Example 3.37 a while-loop inside a for-loop

```
i = 0
for i in range(1, 10):
    j = 1
    while (j <= 10):
        print(i," x ", j," = ", i * j)
        j += 1
    print("Multiplication table for ",i," completed")
```

Example 3.38 while loop inside another while loop

```
i = 0
while(i <= 10):
    j = 1
    while (j <= 10):
        print(i," x ", j," = ", i * j)
        j += 1
    print("Multiplication table for ",i," completed")
    i += 1
```

Observe that more or less the concept of a nested loop is the same in all four options. The inner loop executes for each value of the control_variable of the outer loop.

Example 3.39 Write a program to generate Prime Numbers between two numbers.

```
lower = int(input("Enter lower range: "))
upper = int(input("Enter upper range: "))
if(lower < upper):
  print("Prime numbers between ",lower," and ",upper," are:")
  for num in range(lower, upper):
    p = 1
    for i in range(2, num):
      if(num % i) == 0:
        p = 0
        break
    if(p == 1):
      print(num, "is a Prime Number")
```

Output

```
Enter lower range: 3
Enter upper range: 25
Prime numbers between 3 and 25 are:
3 is a Prime Number
5 is a Prime Number
7 is a Prime Number
11 is a Prime Number
13 is a Prime Number
17 is a Prime Number
```

```
19 is a Prime Number
23 is a Prime Number
>>>
```

3.4 LOOPING THROUGH TWO LISTS

In any other language if there are two arrays we cannot access the elements of both arrays simultaneously. We can access the elements of any one array first and then we can access the elements of the second array.

In Python, we have a wonderful tool called "zip" which combines two or more arrays into a single array. The length of the resulting array will be the length of the smallest array. The rest of the items on the bigger list after the length of the smaller list will be omitted.

The general syntax is: *zip(list1, list2, ...).*

Example 3.40

```
List1 = [1,'rat','mat',4,5+5j]
List2 = ['cat',7,'bat',9.2,10,11,12]
for i, j in zip(List1,List2):
  print(i,"\t",j)
```

Observe that the length of List2 is greater than the length of List1. After applying *zip()*, it will create a new list with a length the same as the smaller list that is List1.

With a single for-loop, we can access the items of both lists simultaneously.

Output

```
1        cat
rat      7
mat      bat
4        9.2
(5+5j)   10
>>>
```

3.5 ITERATOR

The Iterator (iter) is a special data structure in Python. It can "iterate" over by using its index starting at 0 and continuing until the last item of the sequence. Python not only supports iterator on sequences data but it also supports non-sequence datatypes (keys of a dictionary, lines of a file, etc) including user-defined objects. An iterator has a special method *next()*. Just like in any looping structure we increment the control variable to access the next value. An iterator accesses the next item by *iterator.__next__()* method. Once all the items are exhausted, the iterator raises a stop exception.

Example 3.41 Accessing elements of a tuple using iterator.

```
MyTup=(1, 'two', 3.0, 4j,"Five")
i = iter(MyTup)
print("The next item in Tuple is" ,i.__next__())
```

Observe that we are having a tuple, *myTuple*. The statement *i = iter(myTuple)* makes *i* an iterator object. When we execute *i.__next__()* for the first time, it will show the first item of the list.

Output

```
The next item in Tuple is 1
>>>
```

In order to get the rest of the items from the tuple we can use *i.__next__()* repeatedly or we can use a looping structure with exception handling. This is because in the iterator there is no mechanism to find the last element in the Tuple. If we try to access an index that is beyond the size of the tuple, an error will be raised.

Example 3.42 Accessing elements of a tuple using iterator.

```
MyTup=(1, 'two', 3.0, 4j,"Five")
i = iter(MyTup)
while True:
  try:
    print("The next item in Tuple is ",i.__next__())
  except StopIteration:
  break
```

Observe that we have used a while-loop with a try-except mechanism to get all the elements of the tuple.

Output

```
The next item in Tuple is 1
The next item in Tuple is two
The next item in Tuple is 3.0
The next item in Tuple is 4j
The next item in Tuple is Five
>>>
```

It is not possible to move backward, go back to the beginning or copy an iterator. If we want to iterate over the same objects again (or simultaneously), then another iterator object needs to be used.

WORKED OUT EXAMPLES

Example 3.43 Write a program which asks the user to enter an Integer "n" which would be the total number of hours the user worked in a week and calculates and prints the total amount of money the user made during that week. If the user enters any number less than 0 or greater than 168 (n < 0 or n > 168) then your program should print INVALID. Assume that the hourly rate for the first 40 hours is Rs.10 per hour. Hourly rate for extra hours between 41 and 50 (41 <= n <= 50) is Rs.12 per hour. Hourly rate for extra hours greater than 50 is Rs.14 per hour.

```
n = int(input("Enter a positive number :"))
amount = 0
if(n >= 0 and n <= 40):
  extra = n -40
  amount = amount + rate * extra
  print("You made", amount,"rupees this week")
elif(n > 40 and n <= 50):
  extra = n -40
  rate = 12
  amount = 320 + rate * extra
  print("You made", amount,"rupees this week")
elif(n > 50 and n <= 168):
  extra = n -50
  rate = 14
  amount = 410 + rate * extra
  print("You made", amount,"rupees this week")
else:
  print("Invalid")
```

Example 3.44 Write a program that asks the user for a positive number "n" as input. Assume that the user enters a number greater than or equal to 3 and print a triangle as described below. For example if the user enters 3 then the output should be:

```
        *
        **
        ***
        **
        *
```

```
n = int(input("How many Stars: "))

for i in range(1,n+1):
  str1=''
  for j in range(i):
    str1 = str1 + "*"
```

```
    print(str1)

for i in range(n-1, 0, -1):
    str1 = ""
    for j in range(i):
        str1 = str1 + "*"
    print(str1)
```

Example 3.45 Write a program that receives a positive Integer prints "Perfect Number" if the Integer is a perfect number and, "Not a Perfect Number" otherwise.

```
n = int(input("Enter a Number: "))
s = 0
result = "Not a Perfect Number"
for i in range(1, n):
    if(n % i == 0):
        s = s + i
    if(s == n):
        Result = "Perfect Number"
print(n, result)
```

MULTIPLE CHOICE QUESTIONS

1. Find the output of the following code
   x = 20
   if True:
   x = x + 10
   if x == 20:
   x = x + 30
   else:
   x = x + 40
   print(x)

   a. 20
   b. 30
   c. 40
   d. 70

2. Find the output of the following code
   i = 0
   if (i == i):
   i = i
   print(bool(i))

   a. True
   b. False
   c. 0
   d. 1

3. Find the operator that checks if two values are equal

    a. =

    b. ==

    c. ===

    d. !=

4. Find the output of the following code

```
x = 10
y = 0
z = y < x and x > y or y > x and x < y
print(z)
```

    a. True

    b. False

    c. TRUE

    d. FALSE

5. Find the output of the following code

```
x = 10
if x > 5:
x = x + 5
if x < 12:
x = x + 5
if x == 15:
X = x +5
print(x)
```

    a. 10

    b. 15

    c. 20

    d. 25

6. Find the output of the following code

```
i = 0
while i < 3:
i += 2
print("+")
```

    a. 0

    b. 1

    c. 2

    d. 3

7. Find the output of the following code

```
x = 4
if "z" in "computer science":
x = x + 10
elif 5 % 3 == 2:
x = x + 18
```

```
elif 5 > 4:
x = x + 30
else:
x = x + 5
print(x)
```

   a. 20
   b. 21
   c. 22
   d. 23

8. Find how many times "+" will be printed in the output of the following code

```
i = 0
while i <= 5:
i+ = 1
if i % 2 == 0:
break
print("+")
```

   a. 0
   b. 1
   c. 2
   d. 3

9. Find the output of the following code

```
for i in range(1):
print("*")
else:
print("$")
```

   a. One * only
   b. Two * only
   c. One * and a $
   d. Two * and a $

10. Find how many $s will be printed

```
I = 0
while i < 6:
i+= 1
if i % 2 == 0:
continue
print("$")
```

   a. One
   b. Two
   c. Three
   d. Four

11. Find how many $s will be printed
```
i = 0
while i < 10:
print("$")
i = i << 2
```

   a. Two
   b. Four
   c. Eight
   d. Infinite

12. Find how many $s will be printed
```
i = 1
while i < 10:
print("$")
i = i << 1
```

   a. Two
   b. Four
   c. Eight
   d. Infinite

13. Find the output of the following code
```
count = 10
for x in range(0,7):
count = count + 2
if x == 4:
break
print(count)
```

   a. 16
   b. 18
   c. 20
   d. 22

14. Find the output of the following code
```
my_list = ["pet", "dog", 35, "cat", 23]
count = 0
for item in my_list:
if type(item) == str:
continue
count = count +1
print(count)
```

   a. 1
   b. 3
   c. 4
   d. 2

15. Find the output of the following code

```
m = 0
my_str = "mississipi"
for char in my_str:
if char == "s":
continue
if char == "p":
break
m = m + 1
print(m)
```

a. 1
b. 2
c. 3
d. 4

16. Find the output of the following code

```
m = 0
for x in range(4,6):
for y in range(2,4):
m = m + x + y
print(m)
```

a. 25
b. 26
c. 27
d. 28

17. Find the output of the following code

```
m = 0
my_list_1 = [1, 2, 5]
my_list_2 = [1, 3, 2, 6, 5]
for x in my_list_1:
for y in my_list_2:
if x == y:
m = m + 1
print(m)
```

a. 2
b. 3
c. 4
d. 5

18. Find the output of the following code

```
m = 0
my_str_1 = "cat"
my_str_2 = "pet"
for char_1 in my_str_1:
for char_2 in my_str_2:
```

```
if char_1 != char_2:
m = m + 2
print(m)
```

   a. 16
   b. 17
   c. 18
   d. 19

19. Find the output of the following code
```
sum = 0
x = [i**2 for i in range(3)]
l = len(x)
for i in range(l):
sum += x[i]
print(sum)
```

   a. 2
   b. 3
   c. 4
   d. 5

20. Find the output of the following code
```
sum = 0
x = [ 5+i for i in range(2, 7, 3)]
l = len(x)
for i in range(l):
sum += x[i]
print(sum)
```

   a. 15
   b. 16
   c. 17
   d. 18

DESCRIPTIVE QUESTIONS

1. What are the differences between the continue and break statements. Explain with an example.
2. What is a nested if statement? Write how to avoid it.
3. What are the differences between while-else and for-else statements?
4. Explain how range function is used to produce arithmetic progression.
5. What are the use pass statements?
6. What is an iterator? Explain with an example.

PROGRAMMING QUESTIONS

1. Write a program which asks the user to enter their age in years (Assume that the user always enters an Integer) and based on the following conditions,

prints the output: When age is less than or equal to 0, your program should print "UNBORN". When age is greater than 0 and less than or equal to 150, your program should print "ALIVE". When the age is greater than 150, your program should print "LONG LIFE".

2. Write a program which asks the user to enter a positive Integer "n" (Assume that the user always enters a positive Integer) and based on the following conditions, prints the appropriate results:

when "n" is divisible by both 2 and 3 (for example 12), then your program should print "BOTH", when "n" is divisible by only one of the numbers i.e., divisible by 2 but not divisible by 3 (for example 8), or divisible by 3 but not divisible by 2 (for example 9), your program should print "ONE", when "n" is neither divisible by 2 nor divisible by 3 (for example 25), your program should print "NEITHER".

3. Write a program that asks the user for a positive number "n" as input. Assume that the user enters a number greater than or equal to 1 and prints a triangle as described below. For example, if the user enters 4 then the output should be:

```
4
34
234
1234
01234
1234
234
34
4
```

4. Write a Python program to print pattern:

```
* * *
* * * *
* * * * *
* * * *
* * *
```

5. Write a program which reads two numbers and find their GCD.

6. Write a program that will check whether a given number are Armstrong numbers or not. A number is an Armstrong number if the number is an n-digit number, each of its digits raised to the power n is equal to itself.

ANSWER TO MULTIPLE CHOICE QUESTIONS

1.	D	2.	B	3.	B	4.	A	5.	C
6.	C	7.	C	8.	B	9.	C	10.	C
11.	D	12.	B	13.	C	14.	D	15.	D
16.	D	17.	B	18.	A	19.	D	20.	C

4 String

LEARNING OBJECTIVES

After studying this chapter, the reader will be able to:

- Create/initialize a String
- Access elements in a String
- Perform String operations
- Understand character encoding
- Use some common String functions and String methods
- Understand String slicing

4.1 INTRODUCTION

There are several built-in datatypes; some are sequential and some are non-sequential by nature. A sequential datatype allows organizing values and it accesses the data sequentially. It refers to an ordered collection of objects. A non-sequential datatype is an unordered collection and the elements are accessed in a non-sequential manner.

The sequential datatypes are:

1. String.
2. List.
3. Tuple.

The non-sequential datatypes are:

1. Dictionary.
2. Set.

All these datatypes are basically, built-in classes in Python. We will discuss more in regards to classes in Chapter 12.

4.2 STRINGS

A String is used for text manipulation. It is a sequence of characters that may contain one or more combinations of letters, numbers or symbols. A String is immutable in Python. It means once the String is defined, the content of the String cannot be changed/altered. To create a String, we enclose a set of characters within a pair of single quotations ('), double quotations (" ") or triple quotations (''' ''' or """ """).

DOI: 10.1201/9781003219125-5

Example 4.1 Sample strings

```
>>> str1 = "I am Learning Python"
>>> str2 = 'Learning Python is Fun'
>>> print(str1)
I am Learning Python
>>> print(str2)
Learning Python is Fun
>>>
```

In the first case, a String is created with a double quotation and in the second case, the String is created using a single quotation. These quotation marks are not printed. This a just a delimiter to indicate the start and end of the String. If a String includes single or double quotations (to print single/double quotations), we write the String inside a pair of triple quotations.

Example 4.2 Strings that contain single quotations and/or double quotations

```
>>> str1 = """I'm Learning Python"""
>>> str2 = """I told, "He always tell lie" """
>>> print(str1)
I'm Learning Python
>>> print(str2)
I told, "He always tell lie"
>>>
```

Observe that *str1* contains a single quote and *str2* contains a double quote. To print the single quote or double quote, the String *str1* and *str2* are enclosed in pair of triple quotes.

Also, if a String spans more than a line, then also we enclose it in triple quotations.

Example 4.3

```
>>> str1 = """ He told :"Hello! How are you "
I replied: "Fine ! Thank You " """
>>> print(str1)
He told :"Hello! How are you "
I replied: "Fine ! Thank You "
>>>
```

4.2.1 CREATING STRING

There are various ways a String can be created.

4.2.1.1 Creating an Empty String

An *empty string* can be created simply by a pair of single quotes or double quotes with nothing within the quotation marks.

Example 4.4

```
>>> str1 = ' '
>>> print("The empty string with single quotes",str1)
The empty string with single quotes
>>> str2 = " "
>>> print("The empty string with double quotes",str2)
The empty string with double quotes
>>>
```

Observe that here *str1* and *str2* are empty strings.

4.2.1.2   Creating a String from Numbers

We can create a String from a number by type casting *str*.

Example 4.5

```
num = 53
st = str(num)
print("The number in string is ",st)
print("The type of the string is ",type(st))
```

Output

```
The number in string is 53
The type of the string is <class 'str'>
```

4.2.1.3   Creating a String from List and Tuple

We can create a String from a list/tuple that contains non-numeric characters by concatenation operation.

Example 4.6

```
lst1 = ['a','p','p','l','e']
st1 = ""
for ch in lst1:
  st1 = st1 + ch
print("String from list is : ",st1)

tup1 = 'm','a','n','g','o'
st2 = ""
for ch in tup1:
st2 = st2 + ch
print("String from tuple is : ",st2)
```

Observe that *lst1* is a list with few characters and *tup1* is a tuple with some other characters. We can create a String by iterating each item of the list/tuple and concatenating the items. Note that as the concatenation operation is applicable only to strings, it will raise an error if any numeric values are present in the list/tuple.

Output

```
String from list is : apple
String from tuple is : mango
```

4.2.2 ACCESSING STRING

Getting String characters by index refers to accessing String. Python adopts the convention of stating the index of a String with zero. This means that the first character from the left of the String has index 0 and the subsequent character has index 1, 2, 3 and so on as shown in figure 4.1. Unlike any other programming language, Python allows a negative index. The characters in a String can also be accessed starting with the rightmost character of the String. The last character of the String has index –1. The character before last (second last character) has index –2, and so on.

In the String "Hello!", while accessing from left-to-right, the character "H" has index 0 and while accessing from right-to-left it has index –6. While accessing the String elements from left-to-right index starts with 0 and to access the subsequent character we keep on incrementing it by 1, whereas accessing the String elements from right-to-left the index starts with –1 and keeps decrementing by 1.

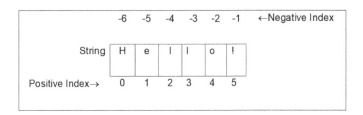

FIGURE 4.1   Accessing String elements using indexing

Example 4.7 Accessing String from left-to-right with positive indices.

```
str1 = "Python is Fun"
length = len(str1)
for i in range(length):
  print(str1[i])
```

Observe that *str1* is a String and each character of the String is accessed by *str1[i]*.

Output

```
P
y
t
h
o
n
```

```
i
s
F
u
n
```

Example 4.8 Accessing list items from right-to-left with negative indices.

```
str1 = "Python is Fun"
length = len(str1)
for i in range(-1, -(length + 1), -1):
  print(str1[i])
```

Output

```
N
u
F
s
i
n
o
h
t
y
P
```

4.2.2.1   Membership in String

The *in* operator determines if a substring is in a String or not. The *not in* operator is the reverse of the *in* operator and determines if an item is not in a sequence.

Example 4.9

```
str1 = "Python is Fun"
str2 = "Python"
str3 = "python"
print(str2 in str1)
print(str3 in str1)
```

Observe that the String *str2* is a substring of *str1*, so the in operator returned True. Whereas *str3* is not a substring in *str1*, so the in operator returned False.

4.2.3   STRING OPERATIONS

For numeric data, we have operations like addition, subtraction, etc. similarly, we have operations that we perform with strings. In this section, we are going to discuss the operations on strings.

4.2.3.1   Concatenation

Strings can be concatenated by using the plus (+) sign. Concatenation means to add a String at the end of the other String.

Example 4.10

```
str1 = "I am LEARNING Python"
str2 = 'Learning Python is Fun'
print(str1 + str2)
z = str1 + str2
print(z)
```

Output

```
I am LEARNING Python Learning Python is Fun
I am LEARNING Python Learning Python is Fun
```

Here we can see str1 is the String "*I am Learning Python*" and str2 is the String '*Learning Python is Fun*'. Either by using *print(str1 + str2)* or by using *z = str1 + str2*. The print gives the same result. It means *str2* is appended to *str1* (i.e., the concatenation of the two strings).

4.2.3.2   Repetition

Strings can be repeated by using the star (*) sign. Repetition means to print the String several times.

Example 4.11

```
str1 = " Python "
z = 3 * str1
print(z)
```

Output

```
Python Python Python
```

Observe that variable str1 contains the String " *Python*". Either by directly printing *3*str1* or by saving it to a variable and then printing gives the same result (i.e., print Python three times in the same line.) It is the same as the concatenation operation applied three times (that is *str1 + str1 + str1*)

Example 4.12

```
str1 = '12'
no = 12
print(5 * str1)
print(5 * no)
```

Output

```
1212121212
60
```

In this example, observe that the value in *str1* is not the same as the value in *no*. The type of the variable *str1* is a String with two characters being 1 and 2. The type of variable *no* is an Integer with the value of it being 12. That is the reason *5*str1* contain1212121212 (i.e., prints 12, five times), while *5*no* results in 60.

Example 4.13

```
str1 = '12'
no = 5
print(str1 + no)
```

Output

```
Traceback (most recent call last):
  File "C:\Users\RakeshN\AppData\Local\Programs\Python\Py
  thon37-32\chap-4.py", line 102, in <module>
  print(str + no)
TypeError: can only concatenate str (not "int") to str1
```

Observe if we try to add str + no, it is obvious that we get an error because we cannot add an Integer to a String. Concatenation (+) is possible when both operands are strings. And addition (+) is possible when both the operands are numeric. To perform addition, we need to convert the String to an Integer by type casting.

Example 4.14

```
Str1 = "12"
no = 12
print(int(str1) + no)
```

Example 4.15 Write a program using while loops that ask the user for a positive Integer "n" and prints a triangle using numbers from 1 to "n".

```
n = int(input("Enter a number :"))
for i in range(1, n + 1):
  print(str(i) * i)
```

Output:

```
Enter a number: 7
1
22
333
```

```
4444
55555
666666
7777777
```

Example 4.16 Write a program that asks the user for an input "n" (assume that the user enters a positive Integer) and prints only the boundaries of the triangle using asterisks "*" of height "n". For example, if the user enters 6 then the height of the triangle should be 6 as shown below and there should be no spaces between the asterisks on the top line:

```
******
*    *
*   *
*  *
* *
**
*
n = 6
if n > 1 :
  print(n * "*") #Prints the TOP line
  for x in range(n - 1, 1, -1) :
    print("*" + (x - 2)*" " + "*")
  print("*")
elif n==1:
  print("*")
```

Output

```
******
*    *
*   *
*  *
* *
**
*
```

4.2.4 CHARACTER ENCODING

Each character is given a unique number or a code, so it can be stored in a computer. This is called *character encoding*. There are many character encoding standards such as ASCII, UTF-8, UTF-16 and UTF-32.

UTF-8 is the main character encoding for the web and simple English text. It can contain any Unicode characters, and more than a million characters can be represented by UTF-8. UTF-8 is using one to four bytes to encode each character. The first 128 characters of UTF-8 are encoded using one byte with the same binary value as ASCII. UTF-8 is recommended as the default encoding for the web pages and in HTML and XML. In UTF-8, **UTF** stands for *Unicode Transformation Format* and **8** stands for *8-bit values that are used for encoding*. table 4.1 shows the most common

TABLE 4.1
Coding in UTF-8

	0	1	2	3	4	5	6	7	8	9	A	B	C	D	E	F
2	SP 0020 *32*	! 0021 *33*	" 0022 *34*	# 002335	$ 0024 *36*	% 0025 *37*	& 0026 *38*	' 0027 *39*	(0028 *40*	) 0029 *41*	* 002A *42*	+ 002B *43*	, 002C *44*	- 002D *45*	. 002E *46*	/ 002F *47*
3	0 0030 *48*	1 0031 *49*	2 0032 *50*	3 0033 *51*	4 0034 *52*	5 0035 *53*	6 0036 *54*	7 0037 *55*	8 0038 *56*	9 0039 *57*	: 003A *58*	; 003B *59*	< 003C *60*	= 003D *61*	> 003E *62*	? 003F *63*
4	@ 0040 *64*	A 0041 *65*	B 0042 *66*	C 0043 *67*	D 0044 *68*	E 0045 *69*	F 0046 *70*	G 0047 *71*	H 0048 *72*	I 0049 *73*	J 004A *74*	K 004B *75*	L 004C *76*	M 004D *77*	N 004E *78*	O 004F *79*
5	P 0050 *80*	Q 0051 *81*	R 0052 *82*	S 0053 *83*	T 0054 *84*	U 0055 *85*	V 0056 *86*	W 0057 *87*	X 0058 *88*	Y 0059 *89*	Z 005A *90*	[005B *91*	\ 005C *92*	] 005D *93*	^ 005E *94*	_ 005F *95*
6	` 0060 *96*	a 0061 *97*	b 0062 *98*	c 0063 *99*	d 0064 *100*	e 0065 *101*	f 0066 *102*	g 0067 *103*	h 0068 *104*	i 0069 *105*	j 006A *106*	k 006B *107*	l 006C *108*	m 006D *109*	n 006E *110*	o 006F *111*
7	p 0070 *112*	q 0071 *113*	r 0072 *114*	s 0073 *115*	t 0074 *116*	u 0075 *117*	v 0076 *118*	w 0077 *119*	x 0078 *120*	y 0079 *121*	z 007A *122*	{ 007B *123*	| 007C *124*	} 007D *125*	~ 007E *126*	DEL 007F *127*

TABLE 4.2
ord and *chr* Functions

Function name	Description
ord ()	takes a character as an argument and returns its UTF-8 Integer value.
chr ()	takes an Integer and returns the UTF-8 character (is the reverse of ord ().)

characters in the first 127 codes in the UTF-8. Notice that this table has 16 columns and 6 rows.

The row's index starts at 2 because the first two rows of this table, which are with the index of 0 and 1, are special non-printable characters. Let us consider the upper-case character P. This character is located in row five and column zero. Its decimal code is 80, which is equivalent to the hexadecimal number 50. Notice that the row of the character is the same as the first hexadecimal digit, and the column of the character is the same as the second hexadecimal digit for that character.

As shown in table 4.2, there are two functions that are useful to convert from a character to a number and from a number to a character.

Example 4.17 Write a program to demonstrate the ord ( ) function.

```
ch = input("Enter any Character : ")
x = ord(ch)
print("The Character encoding with ord() of ",ch," is ",x)
ch1 = chr(x)
print("The Character encoding with chr() of ",x," is ",ch1)
```

Output

```
Enter any Character : P
The Character encoding with ord() of P is 80
The Character encoding with chr() of 80 is P
```

Observe that *ord()* and *chr()* are the opposite of each other. The UTF-8 allows more than a million character codes, and many languages are presented by these character codes.

Example 4.20 Write a program that prints characters between 1000 and 1010

```
for i in range(1000,1010):
  print("The character for ", i, 'is', chr(i))
```

Output

```
The character for 1000 is ϧ
The character for 1001 is ϩ
```

```
The character for 1002 is X
The character for 1003 is x
The character for 1004 is ծ
The character for 1005 is ɞ
The character for 1006 is Ŧ
The character for 1007 is f
The character for 1008 is ӿ
The character for 1009 is Q
```

Sometimes, in character strings, we want to insert a character that is either non-printable, or may have a special meaning. For example, a tab character or a new line is a special character. To access these characters we use a backslash with the character after it. It is called an *escape sequence*. This is another way to present a special character by using an escape sequence. In Python, the use of a backslash is followed by one of the characters as shown in table 4.3.

Example 4.21 Program to demonstrate some escape sequence

```
print("I am Learning Python \nand I am Enjoying it")
print("I am Learning Python \tand I am Enjoying it")
print("I am Learning Python \nand I\'m Enjoying it")
print("I am Learning \"Python\" and I am Enjoying it")
print("I am Learning Python \\ I am Enjoying it")
```

Output

```
I am Learning Python
and I am Enjoying it
I am Learning Python and I am Enjoying it
I am Learning Python
and I'm Enjoying it
I am Learning "Python" and I am Enjoying it
I am Learning Python \ I am Enjoying it
```

TABLE 4.3
Escape Sequences

Escape Sequence	Hex Value	Meaning
\n (New line)	0x0A	We use it to shift the cursor control to the new line.
\t (Horizontal tab)	0x09	We use it to shift the cursor to a couple of spaces to the right in the same line.
\' (Apostrophe or single quotation mark)	0x27	We use it to display the single-quotation mark.
*\" (Double quotation mark)	0x22	We use it to display the double-quotation mark.
\\ (Backslash)	0x5D	We use it to display the backslash character.

4.2.5   STRING FUNCTIONS AND METHODS

A function is applied to data, creates new data and usually returns the produced result. The functions applied to strings are also applied to list, tuple and tuples. Some of the functions are discussed here.

4.2.5.1   len()

This function returns the length of the String. The general syntax is: *len(str)*. The function takes the String name as an argument and returns the number of characters in the String.

> **Example 4.22**
>
> ```
> >>> str1 = "I am Learning Python"
> >>> print("Length of ",str1, "is" , len(str1))
> Length of I am Learning Python is 20
> >>>
> ```

4.2.5.2   del()

This function deletes the entire String. The general syntax is: *del str1*. The function takes the String name as an argument and does not return anything.

> **Example 4.23**
>
> ```
> >>> str1 = "I am Learning Python"
> >>> print(str1)
> I am Learning Python
> >>> del str1
> >>> print(str1)
> Traceback (most recent call last):
>   File "<pyshell#25>", line 1, in <module>
>     print(str1)
> NameError: name 'str1' is not defined
> ```

Observe that when we tried to print after *del str1*, the String is not available. It means the variable name also no more available.

4.2.5.3   max()

This function returns the *maximum* of two String. The general syntax is: *max(String1_name,[String1_name])*. It takes at most two parameters with the second parameter being optional. If only one parameter is passed the *max* function returns the character with the maximum UTF-8 Integer of the String passed. If two parameters are passed, it returns the String in which the character has a maximum UTF-8 Integer when comparing each String character-wise.

Example 4.24

```
st1 = "I am Learning Python"
st2 = "I am Enjoying Python"
print('--',max(st1),'--')
print("Maximum of two strings is ", max(st2,st1))
```

Observe that *y* is the character with the maximum UTF-8 value in *st1*. The second print statement returned *"I am Learning Python"* as while comparing *st1* and *st2*, the first character that is different in these two strings is *"L"* and *"E"*, where *"L"* has a greater UTF-8 value than that of *"E"*.

Output

```
-- y --
Maximum of two strings is I am Learning Python
```

4.2.5.4   min()

This function returns the *minimum* of two strings. The general syntax is: ***min(String1_ name,[String1_name])***. It takes at most two parameters with the second parameter being optional. If only one parameter is passed the *min* function returns the character with the minimum UTF-8 Integer of the String passed. If two parameters are passed, it returns the String in which the first character has a minimum UTF-8 Integer when comparing each String character wise.

Example 4.25

```
st1 = "I am Learning Python"
st2 = "I am Enjoying Python"
print('--',min(st1),'--')
print("Maximum of two strings is ", min(st2,st1))
```

Observe that " " (space) is the character with the minimum UTF-8 value in *st1*. The second print statement returned *"I am Enjoying Python"* as while comparing *st1* and *st2*, the first character that is different in these two strings is *"L"* and *"E"*, where *"E"* has a smaller UTF-8 value than that of *"L"*.

Output

```
-- --
Maximum of two strings is I am Enjoying Python
```

A method is a specific kind of function that behaves like a function and looks like a function, but differs in the way in which it acts, and in its invocation style. A function doesn't belong to any data, rather it gets data. It may create new data and it (generally) produces a result. A method does all these things, but it can also change the state of a selected entity. A method is owned by the data it works for, while a function is owned by the whole code. This also means that invoking a method requires some specification of the data from which the method is invoked. A method is an action which an object can take. In other words, a method is a function that is written for a specific object. Since concepts on object-oriented programming are not yet covered, the only thing that is needed to know at this time is, to use a method, the name of the object to be written, put a dot after it, and then write the name of the method. For example, if the object is a bird, then there are a few actions a bird can do. A fish can swim. In that case, fish is the name of the object, and put a dot, and "can swim" is a method for that object. Remember that for other objects such as a tree, there is no method called swim.

Since methods are like functions, we can also pass some arguments to them. For example, we can say, fish can swim, and in parentheses, pass the variable fast. That means the fish object to swim fast. This concept also applies to objects in programming.

We know a method is a function that is specifically written for a particular object. String methods provide a set of operations that we can perform on a String. Let us discuss some commonly used String methods.

4.2.5.5   capitalize()

This method converts the first character of a given String to its uppercase. The general syntax is: *str.capitalize().*

Example 4.26

```
str1 = "i am learning Python"
str2 = str1.capitalize()
print(str2)
```

Observe that in str, the first character "i" is lowercase and in *"Python"* P is in uppercase. When we apply *str.capitalize()* with no argument, it converts the first character capitalized and the rest of the characters to lowercase.

Output

```
I am learning python
```

4.2.5.6   lower()

This method converts all characters of a given String to its lowercase. The general syntax is: *str.lower().*

Example 4.27

```
str1 = "I AM LEARNING PYTHON"
str2 = str1.lower()
print(str2)
```

Observe that str1 is in upper case. When *str1.lower()* is applied, the String with all the upper cased characters is converted to lowercase.

Output

```
i am learning python
```

4.2.5.7 upper()
This method converts all characters of a given String to its uppercase. The general syntax is: ***str.upper()***.

Example 4.28

```
str1 = "i am learning python"
str2 = str1.upper()
print(str2)
```

Observe that str1 is in lowercase. When *str1.upper()* is applied the String with all the cased characters is converted to uppercase.

Output

```
I AM LEARNING PYTHON
```

4.2.5.8 title()
This method converts the first character of each word of a given String to its uppercase. The general syntax is: ***str.title()***.

Example 4.29

```
str1 = "i am learning python"
str2 = str1.title()
print(str2)
```

Observe that str1 is in lowercase. When *str1.title()* is applied only the first character of each word of a given String is converted to uppercase.

Output

```
I Am Learning Python
```

4.2.5.9 replace()
This method replaces the maximal number of old substrings with a new substring. The general syntax is: ***str.replace(old, new,[count])***. There are three parameters. The first parameter is the existing substring, the second parameter is the new substring with which the old parameter is to be replaced. The third parameter is optional. If the third parameter is absent all the old substrings are replaced. And if the third parameter is present, that is an Integer, replaces that many occurrences form the left.

Example 4.30

```
str1 = "I AM LEARNING PYTHON"
str2 = str1.replace('N', " XXX ")
print(str2)
```

Here the old substring is 'N', the new substring is " XXX " and the third parameter is absent. So, all the occurrences of 'N' is replaced by " XXX ".

Output

```
I AM LEAR XXX I XXX G PYTHO XXX
```

The third parameter is optional. If it is supplied as an Integer, only that many occurrences of the old substring from the left is replaced with the new substring.

Example 4.31

```
str1 = "I AM LEARNING PYTHON"
str2 = str1.replace('N', " XXX ",1)
print(str2)
```

Here we have taken the optional third parameter that is count =1. It means only the first occurrence of the alphabet 'N' is replaced by " XXX ".

Output:

```
I AM LEAR XXX ING PYTHON
```

4.2.5.10   split()

This method splits the given String into a maximal number of words. The general syntax is: ***str.split([chars])***. It may take an optional parameter. If no parameter is passed, It takes the word separator as whitespace by default.

Example 4.32

```
str1 = "I AM LEARNING PYTHON"
str2 = str1.split()
print(str2)
```

Here the *split()* function is used with no parameter. It returned all the words of the input String separated by a comma. In other words, this function takes white space (" ") as a separator.

Output:

```
['I', 'AM', 'LEARNING', 'PYTHON']
```

Example 4.33

```
str1 = "I AM LEARNING PYTHON"
str2 = str1.split('N')
print(str2)
```

Here the parameter passed is 'N'. It means whenever there is an occurrence of character 'N' in the given String, before that character all the characters form a substring. It returns that substring. The character 'N' is omitted.

Output

```
['I AM LEAR', 'I', 'G PYTHO', '']
```

4.2.5.11   strip()

This method removes the leading and trailing characters from a String that is passed as an argument. The general syntax is: *str.strip([chars])*. If the parameter is omitted or if the parameter is None, it removes whitespace. The argument is not a prefix or a suffix; rather, all combinations of its values are stripped.

Example 4.34

```
str1 = "I AM LEARNING PYTHON"
str2 = str1.strip('N')
print(str2)
```

Here in strip(), the optional character 'N' is passed. From the String str1, the last 'N' omitted (as there is no 'N' at the beginning of the String). In other words, if the character is appearing at the beginning or the end of the String, it will be removed.

Output

```
I AM LEARNING PYTHO
```

Similarly, *lstrip()* removes all the leading blank and *rstrip()* removes all the trailing blank of the given String. The reader may try this.

4.2.5.12   count()

This method returns the number of non-overlapping occurrences of a substring in a given String. The general syntax is: *str.count(sub[, start[, end]])*. It takes three parameters. The first parameter is compulsory, it is the substring that we are searching for. The parameter "start" and "end" are optional. If "start" is missing, it is treated as the beginning of the String, if the "end" is missing it is treated as the length of the String.

Example 4.35

```
str1 = "I AM LEARNING PYTHON"
str2 = str1.count('N')
print("Number of \'N\' in the String is ",str2)
```

Here the optional parameter start and end are missing. So *count('N')* returned the number of 'N' that appeared in the given String.

Output

```
Number of 'N' in the String is 3
```

Example 4.36

```
str1 = "I AM LEARNING PYTHON"
str2 = str1.count('N',11)
print("Number of \'N\' from 11th Character onwards is ",str2)
```

Here in this example, *count('N',11)*. It means *start=11*. Starting with the 11th character until the end the function returned the number of N's in the given String.

Output

```
Number of 'N' from 11th Character onwards is 2
```

Example 4.37

```
str1 = "I AM LEARNING PYTHON"
str2 = str1.count('N',3,11)
print("Number of \'N\' from 3rd till 11th Character is ",str2)
```

Here start=3 and end=11, It prints all Ns starting with the 3rd character until the 10th character.

Output

```
Number of 'N' from 3rd till 11th Character is 1
```

4.2.5.13   isalpha()

This method returns True if all characters in the String are alphabetic characters and if there is at least one non-alphabetic character it returns False. The general syntax is: *str.isalpha()*.

Example 4.38

```
str1 = "I AM LEARNING PYTHON"
str2 = str1.isalpha()
print(str2)
```

Here the String *st1* contains blank also. So it returns false.

Output

```
False
```

Example 4.39

```
str1 = "PYTHON"
str2 = str1.isalpha()
print(str2)
```

Here all the characters in st2 are alphabetic. So it returned True.

Output

```
True
```

4.2.5.14   isupper()

This method returns True if all characters in the String are in uppercase and if there is at least one character that is in lowercase, it will return False. The general syntax is: *str.isupper()*.

Example 4.40

```
str1 = "I AM LEARNING PYTHON"
str2 = str1.isupper()
print(str2)
st2 = " I am learning Python"
str2 = st2.isupper()
print(str2)
```

Here *st1* is a String containing all upper case letters. So it returned True. Whereas *st2* is a String containing mixed case alphabets. So it returned False.

Output

```
True
False
```

4.2.6   STRING SLICING

Slicing is a flexible tool to build a new String out of an existing String. Python supports slice notation for any sequential datatype like lists, strings, tuples, bytes, bytearrays and ranges.

Slicing allows extracting or referring to a section of the String, like a slice in a cake. The general syntax is: *string_Name=string[start : end : step]*

Where the *start* is the starting index, the *end* is the ending index, and *step* (optional) is the Integer to skip before getting the next character. Remember that the start index is always less than the *end* index, otherwise, it returns an empty String. And the element in the end index is not included.

Example 4.41

```
str1="I am learning Python"
print("String Before Slicing")
print(str1)
print("\nString After Slicing")
str2=str1[2:10]
print(str2)
```

Here elements from the 2nd index to the 9th index are extracted from the existing String *str1*. Observe that in *str1[2:10]*, the starting index (i.e., 2) is included in str2 whereas the end index (i.e., 10) is not included.

Output

```
String Before Slicing
I am learning Python
String After Slicing
am learn
```

4.2.6.1 String Slicing with Negative Index

Negative indexes and mixed indexes can also be used to slice a list.

Example 4.42

```
str1="I am learning Python"
print("String Before Slicing")
print(str1)
print("\nString After Slicing")
str2=str1[-10:-2]
print(str2)
```

Here the starting index is –10 and the ending index is –2. So items at index –10 and –3 are fetched to str2. Recall that the positive index starts with 0 whereas the negative index starts with –1. It means the character at index –20 is the same as the element at index 0.

Output

```
String Before Slicing
I am learning Python
String After Slicing
ing Pyth
```

Example 4.43

```
str1="I am learning Python"
print("String Before Slicing")
print(str1)
print("\nString After Slicing")
str2=str1[0:-2]
print(str2)
```

Here 0 is the starting index and –2 is the ending index. The end index is excluded from the list.

Output

```
String Before Slicing
I am learning Python
String After Slicing
I am learning Pyth
```

Example 4.44

```
str1="I am learning Python"
print("String Before Slicing")
print(str1)
print("\nString After Slicing")
str2=str1[15:-10]
print(str2)
```

When the starting index is greater than the ending index, it returns an empty String.

Output

```
String Before Slicing
I am learning Python
String After Slicing
```

4.2.6.2 String Slicing with Step

Slicing can be done with a step size also.

Example 4.45 With every alternative element of a given String

```
str1="I am learning Python"
print("String Before Slicing")
print(str1)
length=len(str1)
print("\nString After Slicing")
str2=str1[0:length:2]
print(str2)
```

Here the starting number is 0 and the ending number is the length of the String with *step=2*, so every alternative character of str1 is printed in a forwarded direction.

Output

```
String Before Slicing
I am learning Python
String After Slicing
Ia erigPto
```

Example 4.46 with every alternative element of a given String from in the reverse order.

```
str1="I am learning Python"
print("String Before Slicing")
print(str1)
length=len(str1)
print("\nString After Slicing")
str2=str1[length:0:-2]
print(str2)
```

Here the starting number is the length of the String and the ending number is 0 with *step=−2*, so every alternative character in the String is printed in a reverse direction.

Output

```
String Before Slicing
I am learning Python
String After Slicing
nhy nnalm
```

4.2.6.3   String Slicing Default Index

While slicing a String, the parameters are *start_index*, *end_index* and a *step* value are separated by a colon. If one or more parameters are missing, Python automatically uses the default values.

If the *start_index* is missing, then it is assumed to be zero. If the *end_index* is missing, then Python selects all the way to the end of the list, including the last element. If the step is missing, then it is assumed to be 1.

Example 4.47 Starting_index missing

```
str1="I am learning Python"
print("String Before Slicing")
print(str1)
length=len(str1)
print("\nString After Slicing")
```

```
str2=str1[:-1:2]
print(str2)
```

Here the starting index is missing, so Python takes the default value to be 0.

Output

```
String Before Slicing
I am learning Python
String After Slicing
Ia erigPto
```

Example 4.48 Ending_index missing

```
str1="I am learning Python"
print("String Before Slicing")
print(str1)
length=len(str1)
print("\nString After Slicing")
str2=str1[0::2]
print(str2)
```

Here the *end* index is missing so the default value is the length of the list.

Output

```
String Before Slicing
I am learning Python
String After Slicing
Ia erigPto
```

Example 4.49 step missing

```
str1="I am learning Python"
print("String Before Slicing")
print(str1)
length=len(str1)
print("\nString After Slicing")
str2=str1[0:length:]
print(str2)
```

Here the *step* is missing. Python takes the default step to be 1.

Output

```
String Before Slicing
I am learning Python
String After Slicing
I am learning Python
```

Example 4.50 Starting_index and step missing

```
str1="I am learning Python"
print("String Before Slicing")
print(str1)
print("\nString After Slicing")
str2=str1[:-5:]
print(str2)
```

Here the starting index and the step are missing. The default value for the starting index is 0 and the default value of step is 1.

Output

```
String Before Slicing
I am learning Python
String After Slicing
I am learning P
```

Example 4.51 Ending_index and step missing

```
str1="I am learning Python"
print("String Before Slicing")
print(str1)
print("\nString After Slicing")
str2=str1[4::]
print(str2)
```

Here the ending index and the step are missing. The default value for the ending index is the length of the String (including the last item) and the default value of step is 1.

Output

```
String Before Slicing
I am learning Python
String After Slicing
learning Python
```

Example 4.52 Starting and Ending_index Missing

```
str1="I am learning Python"
print("String Before Slicing")
print(str1)
print("\nString After Slicing")
str2=str1[::2]
print(str2)
```

Here the starting index and the ending index are missing. The default value for the starting index is 0 and the ending index is the length of the String (including the last item).

Output

```
String Before Slicing
I am learning Python
String After Slicing
Ia erigPto
```

Example 4.53 All are missing

```
str1="I am learning Python"
print("String Before Slicing")
print(str1)
print("\nString After Slicing")
str2=str1[::]
print(str2)
```

When starting index, ending index and the step are missing, it is the same as copying a given String to a new String.

Output

```
String Before Slicing
I am learning Python
String After Slicing
I am learning Python
```

WORKED OUT EXAMPLES

Example 4.54 Write a program that accepts an input String consisting of alphabetic characters and removes all the leading whitespace of the String and returns it without using *strip()* function.

```
s = " Hello "
my_index = 0
for x in range(0, len(s)):
  if s[x] != " ":
    break
  new_string = s[x+1 : :]
print(new_string)
```

Example 4.55 Write a program that accepts an input String and returns a String where the case of the characters is changed, i.e. all the uppercase characters are changed to lowercase and all the lowercase characters are changed to uppercase. The non-alphabetic characters should not be changed without using the String methods upper(), lower() or swap().

```
s = 'I am Learning Python'
output_string = ""
for char in s:
```

```
  if (ord (char) <= 90) and (ord (char) >= 65) :
    x = chr (ord (char) + 32)
    output_string += x
  elif (ord (char) <= 122) and (ord (char) >= 97) :
    x = chr (ord (char) -32)
    output_string += x
  else :
    output_string += char
  print (output_string)
```

Example 4.56 Write a program that accepts an input String consisting of alphabetic characters and spaces and returns the String with all the spaces removed without using any String methods.

```
s = ' I Am Learning Python '
out_string = ""
for x in range (0, len (s)) :
  if ( s [x] != " ") :
    out_string = out_string + s [x]
print (out_string)
```

Example 4.57 Given a String of Integers. Which must contain 1 and 2 with the condition that 1 appears before 2. Divide the String into three parts. The first part is the number that appeared before 1, the second part is the number between 1 and 2 (excluding) and the third part is the number after 2. Find the sum of three parts.

For 5417652866 as input, the first part is 54, the second part is 765 and the third part is 866. The sum of these three numbers is 1685.

```
str1 = input ("Enter a String of Integers : ")
index1 = str1.index ('1')
index2 = str1.index ('2')
if (index1 >= index2) :
  print ("1 should appear before 2... Try again... \n")
else:
  num1 = int (str1 [0:index1])
  print ("number before 1: ", num1)
  num2 = int (str1 [index1 + 1 : index2])
  print ("number after 1 and before 2 : ", num2)
  num3 = int (str1 [index2 + 1 : len (str1)])
  print ("num after 2 : ", num3)
  print ("The sum of these three numbers is ", num1 + num2 + num3)
```

MULTIPLE CHOICE QUESTIONS

1. Find the output of the following code
   m = 0
   my_str= "mississipi"

```
for char in my_str:
    if char == "s":
        continue
    if char == "p":
        break
    m = m + 1
print(m)
```

a. 2
b. 3
c. 4
d. 7

2. Find the output of the following code

```
x = "computer science"
print("sci" not in x)
```

a. True
b. TURE
c. False
d. FALSE

3. Find the output of the following code

```
x = "computer science"
del x[2]
print(x)
```

a. coputer science
b. cmputer science
c. empty String
d. Raise an Error

4. Find the output of the following code

```
x = "\\"
print(len(x))
```

a. 0
b. 1
c. 2
d. None of the above

5. Find the output of the following code

```
x = "\"\"\"\"\"\"\"\"\"\"\""
print(len(x))
```

a. 0
b. 1
c. 2
d. None of the above

6. Find the output of the following code

X = """"""
""""""

print(len(x))

   a. 0
   b. 1
   c. 2
   d. None of the above

7. Find the output of the following code

X = '''
'''

print(len(x))

   a. 0
   b. 1
   c. 2
   d. None of the above

8. Find the output of the following code

```
x = "\\\"
print(len(x))
```

   a. 0
   b. 1
   c. 2
   d. Raise an Error

9. Find the output of the following code

```
print(chr(ord('V')+2))
```

   a. U
   b. V
   c. W
   d. X

10. Find the output of the following code

```
x = 5
y = 'Python'
print(y * x)
```

   a. PythonPythonPythonPythonPython
   b. 5*Python
   c. Python*5
   d. Python

11. Find the output of the following code
    s = 'Python is Fun'
    s1 = s[6:–4]
    print(len(s1))

    a. 2
    b. 3
    c. 4
    d. 5

12. Consider the following code
    s = 'Python is Fun'
    s1 = s[6:–4]
    s2 = # Fill this Line
    print(len(s2))
    which code do we have to insert in place of comment to print 2 as an output

    a. s1.lrstrip()
    b. s1.split()
    c. s1.rstrip()
    d. s1.strip()

13. Find the output of the following code
    s = 'AB CD'
    list = list(s)
    list.append('EF')
    print(list)

    a. ['A','B','C','D','EF']
    b. ['A','B',' ','C','D','EF']
    c. ('A','B','C','D','E','F']
    d. {'A','B',' ','C','D','EF'}

14. Consider the following code
    x = 'Vaagdevi'
    y = 'Vaagdevi'
    result = # Fill the Condition
    print(result)
    which statement needs to be substituted from the options given below so that the output will be True?

    a. x < y
    b. x is not y
    c. x is y
    d. x != y

15. Consider the following code
    ```
    a = 10 + 20
    b = '10' + '20'
    c = '10' * 3
    ```
    Identify the types of a, b and c.
    a. a is of int type, b is of int type and c is of int type
    b. a is of int type, b is of str type and c is of str type
    c. a is of int type, b and c are invalid declarations
    d. a is of int type, b is of str type and c is of int type

16. Find the output of the following code
    ```
    s = 'Python is easy'
    s1 = s[-7:]
    s2 = s[-4:]
    print(s1 + s2)
    ```
    a. is easyeasy
    b. is easy easy
    c. easyeasy
    d. iseasyeasy

17. Find the output of the following code
    ```
    s = 'Python is easy'
    s1 = s[-7:-5]
    s2 = s[-4:-2]
    print(s1 + s2)
    ```
    a. isea
    b. is easy
    c. easyeasy
    d. iseasyeasy

18. Find the output of the following code
    ```
    startmsg = 'Python'
    endmsg = ""
    for i in range(0,len(startmsg)):
        endmsg = startmsg[i] + endmsg
    print(endmsg)
        Python
    ```
    a. onhtyP
    b. PyThOn
    c. nohtyP

19. Find the output of the following code
    ```
    str1= "Python Learning is Fun"
    cnt = str1.count('n',3)
    ```

str2 = bool(cnt)
print(str2)

a. TRUE
b. True
c. FALSE
d. False

20. Find the output of the following code
str1= "Python Learning is Fun"
str2 = str1.replace('n',"xxx").count('xx')
print(str2)

a. 4
b. 8
c. 12
d. 16

DESCRIPTIVE QUESTIONS

1. Describe different ways of creating a String. What is the difference between creating a String with double quotes and triple quotes?
2. What are the different String operations? Describe with an example.
3. What is character encoding? What are the different functions available with character encoding?
4. What are different type sequential datatypes?
5. What is an escape sequence?

PROGRAMMING QUESTIONS

1. Write a Python program to get a String from a given String where all occurrences of its last character have been changed to a character given by the user, except the last char itself. Example: input: "reorder" and changed character is '_', output: _eco_de_r
2. Write a Python program to add "ing" at the end of a given String. If the given String already ends with "ing" then add "ly" instead. If the String length of the given String is less than 3, leave it unchanged. Example: input: go, output: go; input: cry, output: crying; input: running, output: runningly.
3. Write a Python program that generates the number of digits in a factorial of a given number. Example: input: 5, output: 3 (factorial of 5 is 120); input: 30, output: 33.
4. Write a Python program that takes a String and prints the longest word in the String. Example: input: The cat says meow, output: 4
5. Write a Python program that checks whether a String is a palindrome or not. Example: input: Madam, output: Palindrome; input: sir, output: Not Palindrome.

6. Write a Python program that changes the cases of a given String. Example: input: Python, output: pYTHON
7. Write a Python program that accepts a String and prints a String after removing a word given by the user. Example: input: The cat says meow, cat, output: The says meow.
8. Write a Python program that counts the number of consonants and vowels in a String. Example: input: The cat says meow, output: 12 no of Consonants and 5 no of vowels.

ANSWER TO MULTIPLE CHOICE QUESTIONS

1.	C	2.	A	3.	D	4.	C	5.	A
6.	A	7.	B	8.	D	9.	D	10.	A
11.	B	12.	D	13.	B	14.	C	15.	B
16.	A	17.	A	18.	D	19.	B	20.	A

5 List and Tuple

LEARNING OBJECTIVES

After Studying this chapter, the reader will be able to:

- Create/initialize a list and a tuple
- Access elements in a list and a tuple
- Perform list and tuple operations
- Use some common list and tuple functions as well as list and tuple methods
- Understand list and tuple slicing

5.1 INTRODUCTION

Sometimes we may have to read, store, process and finally, print hundreds of data. To do this we need to create a separate variable for each value. Handling these variables is a complicated task. To overcome this problem we have one more datatype named "List".

The "Tuple" is yet another data structure which is an immutable collection of objects. When there is a need for an ordered collection of immutable objects, we use tuples.

5.2 LIST

A list is a mutable sequence of objects. It means we can change (update, insert, delete) the objects in a list. It can have data of all different datatypes under one name.

The general syntax for a list is:

Variable_Name = [object 1, object 2 , object 3 , ... , object n]

In a list, notice that:

- A list *starts* and **ends** with a pair of *square brackets* [ ].
- The objects within the square brackets are separated by commas.
- The objects inside a list may have *different types*.
- A list is a collection of **scalar** items.

Example 5.1 Sample strings

```
my_list = [100, "Python", 3.5]
print(my_list)
```

DOI: 10.1201/9781003219125-6

Output

```
[100, "Python", 3.5]
```

We can observe that the list My_list contains different datatypes such as Integer, String, Float, list. We can have other datatypes also.

5.2.1   CREATING LIST

In the previous section, we have discussed the general syntax for creating a list. Here we will learn how to create a list from different data structures.

5.2.1.1   Creating an Empty List

An empty list can be created by simply a pair of square brackets with no data inside.

Example 5.2 Sample strings

```
my_EmptyList = [ ]
print("The list contains ",my_EmptyList)
print("The datatype is ", type(my_EmptyList))
```

Output

```
The list contains [ ]
The datatype is <class 'list'>
```

5.2.1.2   Creating a List from a String

We can create a list from a String by taking each word as an item of the list. This can be done by the *split()* function. Readers are advised to refer to example 4.32 and example 5.33. A list can be created from a String by taking each alphabet of a String also. This is done by *list()* function. The general syntax is: *list(str)*.

Example 5.3

```
str1 = "Python is Fun"
print( "The list is ",list(str1))
```

Output

```
The list is ['P', 'y', 't', 'h', 'o', 'n', ' ', 'i', 's',
' ', 'F', 'u', 'n']
```

5.2.1.3   Creating a List by Range Function

A list can also be created by using the *list* function. We have covered the *range* function in section 3.3.2.1 (readers are advised to refer).

Example 5.4

```
my_list = [ x for x in range(5)]
my_list1 = [ x**3 for x in range(5)]
my_list2 = [ 5 for x in range(5)]
print("The First list is ",my_list)
print("The Second list is ",my_list1)
print("The Third list is ",my_list2)
```

The first statement creates a list of integers from 0 through 4. The second statement creates a list of integers by taking the cube of 0 through 4. And the third statement creates a list of all 5s, 5 times.

Output

```
The First list is [0, 1, 2, 3, 4]
The Second list is [0, 1, 8, 27, 64]
The Third list is [5, 5, 5, 5, 5]
```

5.2.1.4   Creating a List from Another List

A list can be created from a given list by splitting it. This we are going to discuss later in the chapter. One more way of creating a list from another list is by using *list comprehension*. The general syntax is:

[output_Expression_in_ x for x in existing_list [if condition]].

Here the *if condition* is optional.

Example 5.5

```
my_list = [0, 1, 2, 3, 4]
my_list1 = [x*x for x in my_list]
my_list2 = [x**3 for x in my_list if x != 3]
print("The First list is ",my_list1)
print("The Second list is ",my_list2)
```

The second line is optional *if condition* is missing. Here *my_list1* is a new list created from the existing list *my_list* by taking the square of each element of it. The third line uses the optional *if condition*. Here *my_list2* is a new list created from the existing list *my_list* by taking a cube of each element except for 3.

Output

```
The First list is [0, 1, 4, 9, 16]
The Second list is [0, 1, 8, 64]
```

5.2.2 List Operations

5.2.2.1 Concatenation (+)

A new list can also be created from more than one existing list by "+" operation.

Example 5.6

```
my_list = [0, 1, 2, 3, 4]
my_list1 = [5, 6, 7, 8, 9]
my_Newlist = my_list + my_list1
print("The new list is ",my_Newlist)
```

Output

```
The new list is [0, 1, 2, 3, 4, 5, 6, 7, 8, 9]
```

5.2.2.2 Repetition (*)

A single list can be created with repeated objects in it. Repetitions are done by the "*" symbol.

Example 5.7

```
my_list = [1, 2, 3]
my_list1 = my_list * 3
print("The First list is ",my_list)
print("The Second list is ",my_list1)
```

Observe that the list *my_list1* is created by repeating all the elements of the list *my_list* thrice.

Output

```
The First list is [1, 2, 3]
The Second list is [1, 2, 3, 1, 2, 3, 1, 2, 3]
```

5.2.3 Accessing List

To access the list elements, we use an index the same way as we access String elements. Python adopts the convention that the first element in a list is always indexed zero. This means that the item stored at the beginning of the list will have the index number zero or the first element from the left of the list has the index 0. The next element has an index of 1, and so on as shown in figure 5.1.

The value inside the brackets which selects one element from the list is called an *index*. While doing any operation, selecting an element from the list is known as *indexing*. While accessing the list elements from left-to-right the index starts with 0 and keeps on incrementing by 1, whereas accessing the list elements from right-to-left the index starts with –1 and keeps decrementing by 1. This can be seen in

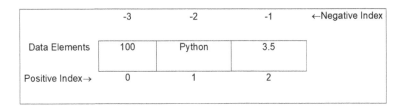

FIGURE 5.1   Accessing list elements using indexing

figure 5.1, the indices and its values while moving from left-to-right are My_list[0] = 100, My_list[1] = "Python" and My_list[2] = 3.5, while the indices and its values while moving from right-to-left are My_list[−3] = 100, My_list[−2] = "Python" and My_list[−1] = 3.5.

Example 5.8 Accessing list items from left-to-right with positive indices

```
my_list = ['crow', 'peacock', 'owl', 'sparrow', 'parrot']
length = len(my_list)
for i in range(length):
  print(my_list[i])
```

Output

```
crow
peacock
owl
sparrow
parrot
```

Example 5.9 Accessing list items from right-to-left with negative indices

```
my_list = ['crow', 'peacock', 'owl', 'sparrow', 'parrot']
length = len(my_list)
for i in range(-1, -(length + 1), -1):
  print(my_list[i])
```

Output

```
parrot
sparrow
owl
peacock
crow
```

5.2.3.1   Membership in List

The *in* operator determines if an item is in a list or not. The *not in* operator is the reverse of the *in* operator and determines if an item is not in a list.

Example 5.10

```
my_list = [1, 2, 3, 4, 5 ]
item = 4
item1 = 8
print (item in my_list)
print (item1 in my_list)
```

Output

```
True
False
```

Observe that the as 4 is an element in my_list it returned True. Whereas 8 is not in my_list, so the in operator returned False.

5.2.4   LIST FUNCTIONS AND METHODS

Just like functions in strings, we have different functions in list also. We are going to discuss about some useful functions.

5.2.4.1.1   len()

This method returns the length of a list. The general syntax is: **len(list)**. The function takes the list's name as an argument and returns the number of elements in the list.

Example 5.11

```
my_list = ['crow', 'peacock', 'owl', 'sparrow', 'parrot']
length = len(my_list)
print("The Length of the list is ",length)
```

Output

```
The Length of the list is 5
```

5.2.4.1.2   del()

The general syntax is: **del(list_name[[start[:end]])**. Apart from the *list_name*, it takes two optional integers as a parameter. If there is no parameter passed, it deletes the entire list. If only one Integer is passed inside a pair of square brackets, it deletes the item at that index. And if two integers are passed separated by a colon inside a pair of square brackets, it deletes all the items between the *start* and *end-1* index. For that, the start index has to be less than the end index. A pair of parentheses is optional.

There is a difference in using the *del()* function while using it for String and for list. While using *del()* in a String, the whole String is deleted. But while using it for a list we can delete the whole list also we can delete a part of the list.

Example 5.12 Deleting the entire list

```
my_list = ['crow', 'peacock', 'owl', 'sparrow', 'parrot']
del(my_list)
print("The list is ",my_list)
```

The output shows that the list does not even exist as the list is deleted.

Output

```
Traceback (most recent call last):
  File "C:\Users\RakeshN\AppData\Local\Programs\Python\Py
  thon37-32\chap-5.py", line 70, in <module>
    print("The list is ",my_list)
NameError: name 'my_list' is not defined
```

Example 5.13 Deleting the element at ith index of the list

```
my_list = ['crow', 'peacock', 'owl', 'sparrow', 'parrot']
print("The list before delete is ",my_list)
del(my_list[2])
print("The list after delete is ",my_list)
```

Here *del my_list[2]* deletes the element at the 2nd index (i.e., owl).

Output

```
The list before delete is ['crow', 'peacock', 'owl',
'sparrow', 'parrot']
The list after delete is ['crow', 'peacock', 'sparrow', 'parrot']
```

This command can delete all the items between the start and end (excluding end).

Example 5.14 Deleting all the elements from start to (end –1) index of the list

```
my_list = ['crow', 'peacock', 'owl', 'sparrow', 'parrot']
print("The list before delete is ",my_list)
del(my_list[1:4])
print("The list after delete is ",my_list)
```

Here all the elements at the 1st, 2nd and 3rd index (i.e., 'peacock', 'owl' and 'sparrow') are deleted.

Output

```
The list before delete is ['crow', 'peacock', 'owl',
'sparrow', 'parrot']
The list after delete is ['crow', 'parrot']
```

5.2.4.1.3   max()

This function returns the *maximum*. A maximum can be found if the list contains objects of the same datatype. It will raise an error if the objects in the list contain mixed datatypes. If the objects in the list are numeric, it will return the maximum number and if the objects in the list are String type, it will return the first element in reverse lexicographical order. The general syntax is: ***max(List_name1,[List_name2])***. This function takes two lists as parameters with the second parameter being optional. If only one parameter is passed it finds the maximum element in the list and if two parameters are passed it returns the maximum of two lists.

Example 5.15 Finding Maximum in a list

```
lst1 = ['crow', 'peacock', 'owl', 'sparrow']
print("List1 is ",lst1)
lst2 = [1,5,2,8, 200,4]
print("List2 is ",lst2)
mx1 = max(lst1)
mx2 = max(lst2)
print("The maximum in list1 is ",mx1," and maximum in list2
is ",mx2)
```

Output

```
List1 is ['crow', 'peacock', 'owl', 'sparrow']
List2 is [1, 5, 2, 8, 200, 4]
The maximum in list1 is sparrow and the maximum in list2 is 200
```

When two lists of the same datatypes are passed as a parameter, this function returns the list which contains the maximum element. It will compare both lists element-wise starting with index 0 and returns the list which has a greater element at a particular index.

Example 5.16 Finding maximum in a list

```
lst1 = [1, 5, 2]
lst2 = [1, 5, 8]
lst3 = [1, 5, 8, 2]
print("Maximum between list1 and list2 ",max(lst1,lst2))
print("Maximum between list2 and list3 ",max(lst2,lst3))
```

Observe that in *lst1* and *lst2* has the same number of elements but the third element is different. The first print statement returned the list which has a greater element at the 3rd position. The second print statement returned *lst3* as it has more elements compared to *lst2*.

Output

```
Maximum between list1 and list2 [1, 5, 8]
Maximum between list2 and list3 [1, 5, 8, 2]
```

5.2.4.1.4 min()

This function returns the *minimum* element. A minimum can be found if the list contains objects of the same datatype. It will raise an error if the objects in the list contain mixed datatypes. If the objects in the list are numeric, it will return the smallest number and if the objects in the list are String type, it will return the first element in lexicographical order. The general syntax is: *min(List_name1,[List_name2])*. This function takes two lists as parameters with the second parameter being optional. If only one parameter is passed it finds the minimum element in the list and if two parameters are passed it returns the minimum of two lists.

Example 5.17 Finding minimum in a list

```
lst1 = ['crow', 'peacock', 'owl', 'sparrow']
print("List1 is ",lst1)
lst2 = [1,5,2,8, 200,4]
print("List2 is ",lst2)
mn1 = min(lst1)
mn2 = min(lst2)
print("The minimum in list1 is ",mn1," and minimum is list2
is ",mn2)
```

Output

```
List1 is ['crow', 'peacock', 'owl', 'sparrow']
List2 is [1, 5, 2, 8, 200, 4]
The minimum in list1 is crow and minimum is list2 is 1
```

There are many in-built methods available in python that are applied to the list. Given below are few commonly used list methods with examples.

5.2.4.1.5 append()

This method adds an item to the end of the list. The general syntax is: *list.appe nd(item)*. The method takes a parameter that is added to an existing list at the end.

Example 5.18 Appending an element into a list

```
my_list = ['crow', 'peacock', 'sparrow', 'parrot']
print("The List before append is ",my_list)
my_list.append('pigeon')
print("The List after append is ",my_list)
```

Here the parameter passed is 'pigeon'. It can be observed that 'pigeon' is added to the existing list as the last element.

Output

```
The List before append is ['crow', 'peacock', 'sparrow', 'parrot']
```

```
The List after append is ['crow', 'peacock', 'sparrow',
'parrot', 'pigeon']
```

5.2.4.1.6   insert()

This method inserts an item at a given position. The general syntax is: *list.insert(i, x)*. The function takes two arguments. The first argument is the index at which the item is inserted. And the second parameter is the item to be inserted.

Example 5.19 Inserting an element into a list

```
my_list = ['crow', 'peacock', 'sparrow', 'parrot']
print("The List before inserting is ",my_list)
my_list.insert(2,'pigeon')
print("The List after insert is ",my_list)
```

In this example, the first parameter is 2. And the second parameter is 'pigeon'. As a result, the item 'pigeon' is inserted to *my_list* in the 2nd location. Remember that the 1st location is the 0th index. In this example, a String is passed as a second parameter.

Output

```
The List before inserting is ['crow', 'peacock', 'sparrow',
'parrot']
The List after insert is ['crow', 'peacock', 'pigeon',
'sparrow', 'parrot']
```

Example 5.20 Inserting an element into a list

```
my_list = ['crow', 'peacock', 'sparrow', 'parrot']
my_Newlist = ['cat', 200]
my_list.insert(2,my_Newlist )
print("my_list after inserting is ",my_list)
```

Observe that the second argument is another list. This list is inserted in the original list at the 2nd index.

Output

```
my_list after inserting is ['crow', 'peacock', ['cat', 200],
'sparrow', 'parrot']
```

5.2.4.1.7   pop()

This method removes the item which is at a given location and returns it. The general syntax is: *list.pop([i])*. It takes an optional parameter. If no index is specified, it removes the last item.

Example 5.21 popping an element from a list

```
my_list = ['crow', 'peacock', 'sparrow', 'parrot', 'owl']
print("The list before pop is ",my_list)
my_list.pop(3)
print("The list after pop is ",my_list)
```

When the *my_list.pop(3)* is called, the element which is in the 3rd index (i.e., 'parrot') is deleted.

Output

```
The list before pop is ['crow', 'peacock', 'sparrow',
'parrot', 'owl']
The list after pop is ['crow', 'peacock', 'sparrow', 'owl']
```

Example 5.22 Popping an element from a list

```
my_list = ['crow', 'peacock', 'sparrow', 'parrot', 'owl']
print("The list before pop is ",my_list)
my_list.pop()
print("The list after pop is ",my_list)
```

If the parameter is not passed, *pop()* deletes the last element of the list.

Output

```
The list before pop is ['crow', 'peacock', 'sparrow',
'parrot', 'owl']
The list after pop is ['crow', 'peacock', 'sparrow',
'parrot']
```

5.2.4.1.8   remove()

This method searches the item in the list and removes it if it is found. If the item appears more than once, the first occurrence item is only removed. The general syntax is: *list.remove(x)*. It takes a single parameter which is the item in the list. The same item may be repeated many times in the list. But the first item from the list whose value is x is deleted.

Example 5.23 Removing an element from a list

```
my_list = ['crow', 'peacock', 'owl', 'sparrow', 'parrot',
'owl']
print("The list before remove is ",my_list)
my_list.remove('owl')
print("The list after remove is ",my_list)
```

Observe that there are two occurrences of 'owl' *in my_list* at indexes 2 and 5. The statement *my_list.remove('owl')* removed the first occurrence of 'owl' which is at index 2.

Output

```
The list before remove is ['crow', 'peacock', 'owl',
'sparrow', 'parrot', 'owl']
The list after remove is ['crow', 'peacock', 'sparrow',
'parrot', 'owl']
```

The difference between *pop(), del* and *remove()* is that *pop()* is the only one which returns the value, and *remove()* is the only one which searches the object, while *del* limits itself to a simple deletion.

5.2.4.1.9   extend()

This method extends a list by appending all its items into a given list. The general syntax is: ***list.extend(L)***. The function takes a single parameter in the form of a list or a String. The original list will not exist once it is extended. If the parameter is a String, every character of the String is treated as an element of the list.

Example 5.24

```
my_list = ['crow', 'peacock', 'owl', 'sparrow', 'parrot']
my_Newlist = ['cat']
my_list.extend(my_Newlist)
print("my_list is ",my_list)
print("my_Newlist is ",my_Newlist)
```

By applying *my_list.extend(my_newList)*, all the elements in *my_newList* are appended to *my_list*. The original *my_list* does not exist anymore. The length of the original list becomes the sum of the length of the original list and the length of the list appended.

Output

```
my_list is ['crow', 'peacock', 'owl', 'sparrow', 'parrot',
'cat']
my_Newlist is ['cat']
```

Example 5.25

```
my_list = ['crow', 'peacock', 'owl', 'sparrow', 'parrot']
my_list.extend('cat')
print("my_list is ",my_list)
```

Observe *list.extend(L)*, takes a single argument of a String. When a String is passed as a parameter, it extends the list by adding each character of the String to it.

Output

```
my_list is ['crow', 'peacock', 'owl', 'sparrow', 'parrot',
'c', 'a', 't']
```

5.2.4.1.10   index()

This method returns the location of the first occurrence of the searched item in the list. The general syntax is: *list.index(x)*. It raises an error if there is no such item available.

Example 5.26

```
my_list = ['crow', 'peacock', 'owl', 'sparrow', 'parrot']
print("The index of sparrow is ",my_list.index('sparrow'))
```

Here the location of 'sparrow' is returned.

Output

```
The index of sparrow is 3
```

5.2.4.1.11   count()

This method returns the number of times the item appears in the list. The general syntax is: *list.count(x)*.

Example 5.27

```
my_list = ['crow', 'peacock', 'owl', 'sparrow', 'parrot']
print("Number times Sparrow appeared in the list : ",
my_list.count('sparrow'))
```

Observe that 'sparrow' appeared only once in the list.

Output

```
Number times Sparrow appeared in the list : 1
```

Example 5.28

```
my_list = ['crow', 'peacock', 'owl', 'sparrow', 'parrot',
'owl']
print("Number times owl appeared in the list : ",my_list
.count('owl'))
```

Here 'owl' appeared twice in the list.

Output

```
Number times owl appeared in the list : 2
```

5.2.4.1.12 sort()

This method sorts the list items. The general syntax is: *list.sort reverse=True|False, key=key)*. It takes two optional parameters. The first parameter is *reverse=True or False* and the second parameter *key* is used for sort customization. If the first parameter *reverse = False,* it returns a sorted list. If the first parameter *reverse = True*, it returns a sorted list in reverse order. If both parameters are absent, it returns a sorted list. If the list contains items of mixed datatype (like a combination of integers and strings), the sort method cannot be applied. Sort method is applied to items of a single datatype.

Example 5.29

```
my_list = ['crow', 'peacock', 'owl', 'sparrow', 'parrot']
print("Unsorted list is : ",my_list)
my_list.sort()
print("Sorted list is : ",my_list)
```

Here both parameters are absent. It returns a sorted list.

Output

```
Unsorted list is : ['crow', 'peacock', 'owl', 'sparrow',
'parrot']
Sorted list is : ['crow', 'owl', 'parrot', 'peacock',
'sparrow']
```

Example 5.30

```
my_list = ['crow', 'peacock', 'owl', 'sparrow', 'parrot']
print("Unsorted list is : ",my_list)
my_list.sort(reverse = True)
print("Reversely Sorted list is : ",my_list)
```

Here the parameter is *reverse = True*. It returned a sorted list in reverse order.

Output

```
Unsorted list is : ['crow', 'peacock', 'owl', 'sparrow', 'parrot']
Reversely Sorted list is : ['sparrow', 'peacock', 'parrot',
'owl', 'crow']
```

Example 5.31

```
my_list = ['crow', 'peacock', 'owl', 'sparrow', 'parrot']
print("Unsorted list is : ",my_list)
```

```
my_list.sort(key = len)
print("Sorted list based on length of item is : ",my_list)
```

Observed that in this example, the parameter *key=len* is passed. Here *len* is the function that finds the length of a String. It means the items are sorted as per the length of the list. Here the sorting is done based on the length of the String. We can also pass the name of the user-defined function as a parameter.

Output

```
Unsorted list is : ['crow', 'peacock', 'owl', 'sparrow',
'parrot']
Sorted list based on length of item is : ['owl', 'crow',
'parrot', 'peacock', 'sparrow']
```

The value of the key parameter should be a function that takes a single argument and returns a key to use for sorting purposes.

Example 5.32

```
my_list = ['crow', 'peacock', 'owl', 'sparrow', 'parrot']
print("Unsorted list is : ",my_list)
my_list.sort(reverse = True, key = len)
print("Reversely Sorted list based on length of item is :
",my_list)
```

Here observe that both the parameters are present. It sorts the list according to the length in reverse order.

Output

```
Unsorted list is : ['crow', 'peacock', 'owl', 'sparrow', 'parrot']
Reversely Sorted list based on length of item is :
['peacock', 'sparrow', 'parrot', 'crow', 'owl']
```

5.2.4.1.13   sorted()

This function also sorts the list items. The general syntax is: *sorted(list, key=key, reverse=True|False)*. It takes three parameters, the first one is the list_name on which the sorting is to be done and it is compulsory. The other two parameters are optional. The second parameter *key* is used for sort customization. The third parameter *reverse=True* refers to sorting the list in reverse order. When only one parameter, that is the list_name is passed, it sorts the given list.

Example 5.33

```
my_list = ['crow', 'peacock', 'owl', 'sparrow', 'parrot']
print("Unsorted list is : ",my_list)
my_Newlist = sorted(my_list)
print("Sorted list is : ",my_Newlist)
```

Output

```
Unsorted list is : ['crow', 'peacock', 'owl', 'sparrow', 'parrot']
Sorted list is : ['crow', 'owl', 'parrot', 'peacock', 'sparrow']
```

When two parameters are passed that is along with the list_name, the key is also passed and it sorts the list as per the key.

Example 5.33

```
my_list = ['crow', 'peacock', 'owl', 'sparrow', 'parrot']
print("Unsorted list is : ",my_list)
my_Newlist = sorted(my_list, key = len)
print("Sorted list is : ",my_Newlist)
```

Observe that here the key is the String length. It returns a sorted list based on the length of the items.

Output

```
Unsorted list is : ['crow', 'peacock', 'owl', 'sparrow', 'parrot']
Sorted list is : ['owl', 'crow', 'parrot', 'peacock', 'sparrow']
```

When the third parameter is also passed, it can generate a list that is sorted in reverse order.

Example 5.34

```
my_list = ['crow', 'peacock', 'owl', 'sparrow', 'parrot']
print("Unsorted list is : ",my_list)
my_Newlist = sorted(my_list, key = len, reverse = True)
print("Sorted list is : ",my_Newlist)
```

Output

```
Unsorted list is : ['crow', 'peacock', 'owl', 'sparrow', 'parrot']
Sorted list is : ['peacock', 'sparrow', 'parrot', 'crow', 'owl']
```

The main difference between *sort()* and *sorted()* is:

1. After using the *sort()* method on a list, the original list is lost. But *sorted()* function creates a new list which is sorted.
2. After using *sorted()*, we have two lists, the original list and a sorted list whereas after using *sort()* we have only one list, which is passed.
3. If we want to have the unsorted version of the list even after sorting, we use *sorted()*.
4. The *sorted()* function will return a sorted list whereas the *sort()* method is an in-place sorting algorithm, it does not return anything.

5.2.4.1.14   reverse()
This method reverses the elements of the list in place. The general syntax is: *list. reverse()*. It takes an optional parameter.

Example 5.34
```
my_list = ['crow', 'peacock', 'owl', 'sparrow', 'parrot']
print("Original list is : ",my_list)
my_list.reverse()
print("Reversed list is : ",my_list)
```

Output
```
Original list is : ['crow', 'peacock', 'owl', 'sparrow',
'parrot']
Reversed list is : ['parrot', 'sparrow', 'owl', 'peacock',
'crow']
```

5.2.4.1.15   copy()
This method returns a copy of the given list. The general syntax is: *list.copy()*. It creates a new and independent list with different references. When a copy is created with an equal sign, it has the same reference.

Example 5.35
```
my_list = ['crow', 'peacock', 'owl', 'sparrow', 'parrot']
print("Original list is : ",my_list)
print("Id of the original list is : ", id(my_list))
my_Newlist = my_list.copy()
print("The copy of the original list is : ", my_Newlist)
print("Id of the new list is : ", id(my_Newlist))
my_Newlist1 = my_list
print("The copy with \'=\' is : ", my_Newlist1)
print("Id of the copy with \'=\' is : ", id(my_Newlist1))
```

Here *my_list, my_Newlist* and *my_Newlist1* contain the same data. But observe that the *id* of *my_list* is the same as the *id* of *my_Newlist1* whereas the *id* of *my_list* and the *id* of *my_Newlist* are different. *my_Newlist1* is created by using an "=" symbol (equal to operator) whereas *my_Newlist* is created by using the *copy()* method.

Output
```
Original list is : ['crow', 'peacock', 'owl', 'sparrow',
'parrot']
Id of the original list is : 64272712
```

```
The copy of the original list is : ['crow', 'peacock',
'owl', 'sparrow', 'parrot']
Id of the new list is : 29548680
The copy with '=' is : ['crow', 'peacock', 'owl', 'sparrow',
'parrot']
Id of the copy with '=' is : 64272712
```

5.2.4.1.16   clear()

The method removes all items from the list. The general syntax is: ***list.clear()***. It is equivalent to del a[:]. The *clear()* function deletes all the items from the given list. It means there is an empty list existing after the operation.

Example 5.36

```
my_list = ['crow', 'peacock', 'owl', 'sparrow', 'parrot']
my_Newlist = my_list.clear()
print("The list after CLEAR :",my_Newlist)
```

Output

```
The list after CLEAR : None
```

5.2.4.1.17   reversed()

This function returns an iterator that traverses a control variable in reverse order. The general syntax is: ***list(reversed(list))***.

Example 5.37

```
my_list = [1,'two', 3.0,4j, True]
print("The Original list is :",my_list)
my_Newlist = list(reversed(my_list))
print("The revesed list is :",my_Newlist)
```

Output

```
The Original list is : [1, 'two', 3.0, 4j, True]
The revesed list is : [True, 4j, 3.0, 'two', 1]
```

5.2.4.1.18   enumerate()

This function adds a counter to a control variable and returns it in a form of enumerating object. The general syntax is: ***list(enumerate(list),[start=])***. It takes two parameters. The first parameter is the *list_name*. The second parameter is an optional Integer. If the second parameter is absent, the enumeration stats with zero by default. Otherwise, the enumeration starts with the Integer given in the second parameter.

Example 5.38

```
my_list = [1,'two', 3.0,4j, True]
print("The Original list is :",my_list)
my_Newlist = list(enumerate(my_list))
print("The enumerated list is :",my_Newlist)
```

Observe that the enumeration starts at 0. It means the 0th item in the list is 1, the 1st item is 'two' and so on.

Output

```
The Original list is : [1, 'two', 3.0, 4j, True]
The enumerated list is : [(0, 1), (1, 'two'), (2, 3.0),
(3, 4j), (4, True)]
```

We can change the starting number of enumerations. By passing start=1 as a parameter along with the list.

Example 5.39

```
my_list = [1,'two', 3.0,4j, True]
print("The Original list is :",my_list)
my_Newlist = list(enumerate(my_list,start=1))
print("The enumerated list starting with 1 is :",my_Newlist)
```

Observe that the enumeration started with 1 as we passed start=1. We can have any starting number.

Output

```
The Original list is : [1, 'two', 3.0, 4j, True]
The enumerated list starting with 1 is : [(1, 1), (2,
'two'), (3, 3.0), (4, 4j), (5, True)]
```

5.2.4.1.19   any()

It will return True if any items traversed across an iterator have a Boolean True value. The general syntax is: *any(list)*. We can think of this as a Boolean OR operation. If any of the items in the list represents Boolean True, it will return a True, otherwise, it will return False.

Example 5.40

```
my_list1 = [1, False]
print("The First list is :",my_list1)
print("If any of the item has Boolean values :",
any(my_list1))
my_list2 = [0, False]
```

```
print("The Second list is :",my_list2)
print("If any of the item has Boolean values :",
any(my_list2))
my_list3 = [1, True]
print("The Third list is :",my_list3)
print("If any of the item has Boolean values :",
any(my_list3))
my_list4 = [0, 1]
print("The Fourth list is :",my_list4)
print("If any of the item has Boolean values :",
any(my_list4))
my_list5 = [ ]
print("The Fifth list is :",my_list5)
print("If any of the item has Boolean values :",
any(my_list5))
```

Remember that 0 represents "False" and 1 represents "True". Any numeric value greater than 1 is also taken as True. Observe that in myList1, myList3 and myList4 either we are having truth value True or we have all the values as numeric values. So the output is True. For myList2 and myList5 the output is False as either all the values represent Boolean False or the list is empty.

Output

```
The First list is : [1, False]
If any of the item has Boolean values : True
The Second list is : [0, False]
If any of the item has Boolean values : False
The Third list is : [1, True]
If any of the item has Boolean values : True
The Fourth list is : [0, 1]
If any of the item has Boolean values : True
The Fifth list is : []
If any of the item has Boolean values : False
```

5.2.4.1.20   all()

It will return True if all items traversed across an iterator have a Boolean True value. The general syntax is: *all(list)*. We can think of this as Boolean AND operation. If the entire item in the list represents Boolean True, it will return a True, otherwise, it will return False.

Example 5.41

```
my_list1 = [1, False]
print("The First list is :",my_list1)
print("If all of the item has Boolean values :",
all(my_list1))
my_list2 = [0, False]
print("The Second list is :",my_list2)
```

```
print("If all of the item has Boolean values :",
all(my_list2))
my_list3 = [1, True]
print("The Third list is :",my_list3)
print("If all of the item has Boolean values :",
all(my_list3))
my_list4 = [0, 1]
print("The Fourth list is :",my_list4)
print("If all of the item has Boolean values :",
all(my_list4))
my_list5 = [ ]
print("The Fifth list is :",my_list5)
print("If all of the item has Boolean values :",
all(my_list5))
```

Observe that only myList3 and myList5 return True. It is because either the list is empty or all the values have truth values not representing False. In myList1, myList2 and myList4, at least one value represents False so it is the output.

Output

```
The First list is : [1, False]
If all of the item has Boolean values : False
The Second list is : [0, False]
If all of the item has Boolean values : False
The Third list is : [1, True]
If all of the item has Boolean values : True
The Fourth list is : [0, 1]
If all of the item has Boolean values : False
The Fifth list is : []
If all of the item has Boolean values : True
```

Example 5.42 Write a program that finds the sum of numbers in a list

```
my_list = [1, 2,3,4,5]
sum = 0
for i in my_list:
  sum += i
print("The sum is ", sum )
```

Output

```
The sum is 15
```

Example 5.43 Write a program that asks the user to enter a few numbers until –1. Create a list with all even numbers supplied by the user.

```
i = int(input("Enter a Number : "))
even_list = []
```

```
while i != -1:
  if i % 2 == 0:
    even_list.append(i)
  i = int(input("Enter next number (-1 to end) : "))
print("The list of even numbers is : ",even_list)
```

Output

```
Enter a Number : 4
Enter next number (-1 to end) : 5
Enter next number (-1 to end) : 23
Enter next number (-1 to end) : 44
Enter next number (-1 to end) : 46
Enter next number (-1 to end) : 55
Enter next number (-1 to end) : -1
The list of even numbers is : [4, 44, 46]
```

5.2.5   LIST SLICING

Slicing is a flexible tool to build new lists out of an existing list. Python supports slice notation for any sequential datatype like lists, strings, tuples, bytes, byte-arrays and ranges. To replace an item in a list, simply put the element that we would like to change on the left-hand side of an equal sign, and assign a new value to it.

Example 5.44

```
my_list = ['crow', 'peacock', 'owl', 'sparrow', 'parrot', 34]
print("The Original list is :",my_list)
length = len(my_list)
my_list[length - 1] = 'swan'
print("The Updated list is :",my_list)
```

In this example, we can see the last item in the list is 34. Assume that it is a wrong entry; we are going to replace 34 with 'swan'. This can be done by identifying the index of the wrong entry. And simply write the correct entry in that location. Always remember that while traversing the list from left-to-right use positive indices starting with 0 or while traversing from right-to-left using negative indices starting with –1. The output remains the same.

Example 5.45

```
my_list = ['crow', 'peacock', 'owl', 'sparrow', 'parrot', 34]
print("The Original list is :",my_list)
length = len(my_list)
my_list[-1] = 'swan'
print("The Updated list is :",my_list)
```

Output

```
The Original list is : ['crow', 'peacock', 'owl', 'sparrow',
'parrot', 34]
The Updated list is : ['crow', 'peacock', 'owl', 'sparrow',
'parrot', 'swan']
```

Slicing allows us to extract or refer to a section of the list, like a slice in a cake. The general syntax is: *list_Name = List[start : end : step].* Where the *start* is the starting index, the *end* is the ending index and *step* (optional) is the Integer to skip before getting the next item. Remember that the start index is always less than the end index. And the item in the end index is not included.

Example 5.46

```
my_list = ['crow', 'peacock', 'owl', 'sparrow', 'parrot', 34]
print("The Original list is :",my_list)
length = len(my_list)
my_list = my_list[2:4]
print("The Updated list is :",my_list)
```

Here we have extracted an element, which is at index 2 and 3 and places it in a new list. Observe that in my_list[2:4], the 1st index (i.e., 2) is included in the new list whereas the 2nd index (i.e., 4) is not included in the new list.

Output

```
The Original list is : ['crow', 'peacock', 'owl', 'sparrow',
'parrot', 34]
The Updated list is : ['owl', 'sparrow']
```

From the examples discussed above observe that the item in the starting index is included whereas the item in the ending index is not included while slicing to get a new list.

5.2.5.1   List Slicing with Negative Index

Negative indexes and mixed indexes can also be used to slice a list.

Example 5.47

```
my_list = ['crow', 'peacock', 'owl', 'sparrow', 'parrot', 34]
print("The Original list is :",my_list)
length = len(my_list)
my_list = my_list[-5:-2]
print("The Updated list is :",my_list)
```

Here elements at index –5, –4 and –3 are fetched to the new list. Recall that the positive index starts with 0 whereas the negative index starts with –1. It means an element at index –5 is the same as the element at index 0.

Output

```
The Original list is : ['crow', 'peacock', 'owl', 'sparrow',
'parrot', 34]
The Updated list is : ['peacock', 'owl', 'sparrow']
```

Example 5.48

```
my_list = ['crow', 'peacock', 'owl', 'sparrow',
'parrot', 34]
print("The Original list is :",my_list)
length = len(my_list)
my_list = my_list[0:-2]
print("The Updated list is :",my_list)
```

Here 0 is the starting index and –2 is the ending index. 'crow' is at index 0 and 'sparrow' is at index –2. The end index is excluded from the list. So the output is:

Output

```
The Original list is : ['crow', 'peacock', 'owl', 'sparrow',
'parrot', 34]
The Updated list is : ['crow', 'peacock', 'owl', 'sparrow']
```

Example 5.49

```
my_list = ['crow', 'peacock', 'owl', 'sparrow', 'parrot', 34]
print("The Original list is :",my_list)
length = len(my_list)
my_list = my_list[-4:4]
print("The Updated list is :",my_list)
```

Here the starting index is –4 with data 'peacock' and the end index is 4 with data 'parrot'. So, the new list will contain all the elements starting with 'peacock' and ending at 'sparrow'.

Output

```
The Original list is : ['crow', 'peacock', 'owl', 'sparrow',
'parrot', 34]
The Updated list is : ['owl', 'sparrow']
```

Example 5.50

```
my_list = ['crow', 'peacock', 'owl', 'sparrow', 'parrot', 34]
print("The Original list is :",my_list)
length = len(my_list)
my_list = my_list[2:-2]
print("The Updated list is :",my_list)
```

Output

```
The Original list is : ['crow', 'peacock', 'owl', 'sparrow',
'parrot', 34]
The Updated list is : ['owl', 'sparrow']
```

Example 5.51

```
my_list = ['crow', 'peacock', 'owl', 'sparrow', 'parrot', 34]
print("The Original list is :",my_list)
length = len(my_list)
my_list = my_list[-2:2]
print("The Updated list is :",my_list)
```

Output

```
The Original list is : ['crow', 'peacock', 'owl', 'sparrow',
'parrot', 34]
The Updated list is : []
```

5.2.5.2   List Slicing with Step

Slicing can be used by including a step size also. The format for doing this is the same as the slicing list and adding a step at the end. So to specify the start index, end index and also a step separated by a colon.

Example 5.52 Write a program to generate a new list with every alternative element of a given list

```
my_list = ['crow', 'peacock', 'owl', 'sparrow', 'parrot']
print("The original list is ", my_list)
length = len(my_list)
my_list = my_list[0:length:2]
print("The updated list is ", my_list)
```

Here the starting number is 0 and the ending number is the length of the String with *step = 2,* so every alternative valued in the list is printed in a forwarded direction.

Output

```
The original list is ['crow', 'peacock', 'owl', 'sparrow',
'parrot']
The updated list is ['crow', 'owl', 'parrot']
```

Example 5.53 Write a program to generate a new list with every alternative element of a given list in the reverse order

```
my_list = ['crow', 'peacock', 'owl', 'sparrow', 'parrot']
print("The original list is ", my_list)
```

```
length = len(my_list)
my_list = my_list[length:0:-2]
print("The updated list is ", my_list)
```

Here the starting number is the length of the String and the ending number is 0 with *step = –2*, so every alternative valued in the list is printed in a reverse direction.

Output

```
The original list is ['crow', 'peacock', 'owl', 'sparrow',
'parrot']
The updated list is ['parrot', 'owl']
```

Example 5.54

```
my_list = ['crow', 'peacock', 'owl', 'sparrow', 'parrot']
print("The original list is ", my_list)
my_list = my_list[0:-1:3]
print("The updated list is ", my_list)
```

As the step = 3 so every third element is printed.

Output

```
The original list is ['crow', 'peacock', 'owl', 'sparrow',
'parrot']
The updated list is ['crow', 'sparrow']
```

Example 5.55 Write a program to generate a reverse list except for the first element

```
my_list = ['crow', 'peacock', 'owl', 'sparrow', 'parrot']
print("The original list is ", my_list)
my_list = my_list[-1:0:-1]
print("The updated list is ", my_list)
```

As the ending index is 0, data in this index is excluded and step = –1 indicates reverse order.

Output

```
The original list is ['crow', 'peacock', 'owl', 'sparrow',
'parrot']
The updated list is ['parrot', 'sparrow', 'owl', 'peacock']
```

5.2.5.3   List Slicing Default Index

While slicing a list, the parameters are *start_index*, *end_index* and a step value separated by a colon. If one or more parameters are missing, Python automatically uses the default values. If the start_index is missing, then it is assumed to be zero. If the end_index, is missing, then Python selects all the way to the end of the list, including the last element. If the step is missing, then it is assumed to be 1.

Example 5.56 Starting index is missing

```
my_list = ['crow', 'peacock', 'owl', 'sparrow', 'parrot']
print("The original list is ", my_list)
my_list = my_list[:-1:2]
print("The updated list is ", my_list)
```

Here the starting index is missing, so python takes the default value to be 0.

Output

```
The original list is ['crow', 'peacock', 'owl', 'sparrow',
'parrot']
The updated list is ['crow', 'owl']
```

Example 5.57 The ending index is missing

```
my_list = ['crow', 'peacock', 'owl', 'sparrow', 'parrot']
print("The original list is ", my_list)
my_list = my_list[::2]
print("The updated list is ", my_list)
```

Here the ending index is missing so the default value is the length of the list.

Output

```
The original list is ['crow', 'peacock', 'owl', 'sparrow',
'parrot']
The updated list is ['crow', 'owl', 'parrot']
```

Example 5.58 Step missing

```
My_list = ['crow', 'peacock', 'owl', 'sparrow', 'parrot']
print("The Original list is :",my_list)
my_list = my_list[2:4:]
print("The Updated list is :",my_list)
```

Here the step function is missing. Python takes the default step to be 1.

Output

```
The Original list is : ['crow', 'peacock', 'owl', 'sparrow',
'parrot']
The Updated list is : ['owl', 'sparrow']
```

Example 5.59 Starting_index and step missing

```
my_list = ['crow', 'peacock', 'owl', 'sparrow', 'parrot']
print("The original list is ", my_list)
my_list = my_list[:-1:]
print("The updated list is ", my_list)
```

Here the starting index and the step is missing. The default value for the starting index is 0 and the default value of the step is 1.

Output

```
The original list is ['crow', 'peacock', 'owl', 'sparrow',
'parrot']
The updated list is ['crow', 'peacock', 'owl', 'sparrow']
```

Example 5.60 Ending_index and step missing

```
my_list = ['crow','peacock','owl','sparrow','parrot']
print("The original list is ", my_list)
my_list = my_list[2::]
print("The updated list is ", my_list)
```

Output

```
The original list is ['crow', 'peacock', 'owl', 'sparrow',
'parrot']
The updated list is ['owl', 'sparrow', 'parrot']
```

Example 5.61 Starting and ending index missing

```
my_list = ['crow','peacock','owl','sparrow','parrot']
print("The original list is ", my_list)
my_list = my_list[::2]
print("The updated list is ", my_list)
```

Output

```
The original list is ['crow', 'peacock', 'owl', 'sparrow',
'parrot']
The updated list is ['crow', 'owl', 'parrot']
```

Example 5.62 All the parameters are missing

```
my_list = ['crow','peacock','owl','sparrow','parrot']
print("The original list is ", my_list)
```

```
my_list = my_list[::]
print("The updated list is ", my_list)
```

When starting index, ending index and the step are missing it is the same as copying the given list to a new variable.

Output

```
The original list is ['crow', 'peacock', 'owl', 'sparrow',
'parrot']
The updated list is ['crow', 'peacock', 'owl', 'sparrow',
'parrot']
```

5.2.6   DIFFERENCE BETWEEN ASSIGNMENT AND COPYING A LIST

The assignment is done by using an equal to (=) symbol. When we perform a list assignment, it does not create a separate list, rather it creates a separate variable referring to the same location as the original list. Any change to the first list reflects the change in the second list also.

Example 5.63

```
lst1 = [1,2,3,4,5]
lst2 = lst1
print("Location of lst1 is :", id(lst1))
print("Location of lst2 is :", id(lst2))
lst2.append(6)
print("lst1 after appending :", lst1)
print("lst2 after appending :", lst2)
```

Output

```
Location of lst1 is : 63748456
Location of lst2 is : 63748456
lst1 after appending : [1, 2, 3, 4, 5, 6]
lst2 after appending : [1, 2, 3, 4, 5, 6]
```

We perform the copy operation by using a [:] operation. It creates a separate list with the same data at different memory locations. Any change to one list will not reflect the other list.

Example 5.64

```
lst1 = [1,2,3,4,5]
lst2 = lst1[:]
print("Location of lst1 is :", id(lst1))
print("Location of lst2 is :", id(lst2))
```

```
lst2.append(6)
print("lst1 after appending :", lst1)
print("lst2 after appending :", lst2)
```

Output

```
Location of lst1 is : 63068648
Location of lst2 is : 24633480
lst1 after appending : [1, 2, 3, 4, 5]
lst2 after appending : [1, 2, 3, 4, 5, 6]
```

5.3 MULTI-DIMENSIONAL LISTS

There are situations when it is required to store the data as a multi-dimensional structure. For example, if we have to represent a matrix or we have a small list of employee records. Generally, it is represented in two dimensions-one of the dimensions being rows and the other one to be columns. In Python, multi-dimensional data can be stored as a list of lists. The general syntax is:

Variable_Name = [[item 11, item 12, ..., item 1n], [item 21, item 22, ..., item 2n], ..., [item m1, item m2, ..., item mn]]

As usual, each list is enclosed by a pair of square brackets and each item in every list is separated by a comma. Let us try to understand this with the help of an example. As shown in table 5.1, consider an Employee_Record

Assuming Emp_Id, Emp_Name and Phone_No are String type data while Salary is an Integer type data. There are three records with four items in each record. By using Python, we can represent the Employee_Record in terms of a list as:

Employee_Record= [['501', 'R.Naveen Kumar', '9876543210', 130000], ['502', 'J.Srikanth', '9988334477', 125000], ['503', 'Mahipal Reddy', '8899223377', 60000]]

Observe that each record is kept within a pair of square brackets and each item in a record is separated by a comma.

5.3.1 RETRIEVAL FROM MULTI-DIMENSIONAL LISTS

We already know how to retrieve data from a list by using an index. Similarly, we will retrieve data from a multi-dimensional list using two indices.

TABLE 5.1
Sample Employee Record

Emp_Id	Emp_Name	Phone_No	Salary
501	R.Naveen Kumar	9876543210	130000
502	J.Srikanth	9988334477	125000
503	Mahipal Reddy	8899223377	60000

Example 5.65

```
Employee_Record = [['501', 'R.Naveen Kumar', '9876543210',
130000],
        ['502', 'J.Srikanth', '9988334477', 125000],
        ['503', 'Mahipal Reddy', '8899223377', 60000]]
print(Employee_Record[0])
```

Observe that *Employee_Record[0]* refers to the 1st record of the multi-dimensional list.

Output

```
['501', 'R.Naveen Kumar', '9876543210', 130000]
```

Example 5.66

```
Employee_Record = [['501', 'R.Naveen Kumar', '9876543210',
130000],
        ['502', 'J.Srikanth', '9988334477', 125000],
        ['503', 'Mahipal Reddy', '8899223377', 60000]]
for i in range(len(Employee_Record)):
print(Employee_Record[i])
```

All the records are retrieved by using an index. Observe that we have used a looping-structure to retrieve.

Output

```
['501', 'R.Naveen Kumar', '9876543210', 130000]
['502', 'J.Srikanth', '9988334477', 125000]
['503', 'Mahipal Reddy', '8899223377', 60000]
```

To access the item in a list two indices required. For example, Employee_Record[1][1] will retrieve 'J.Srikanth'.

Example 5.67

```
Employee_Record = [['501', 'R.Naveen Kumar', '9876543210',
130000],
        ['502', 'J.Srikanth', '9988334477', 125000],
        ['503', 'Mahipal Reddy', '8899223377', 60000]]
print(Employee_Record[1][0])
```

Output

```
502
```
We can also retrieve the alphabet of a String which is an item in a list.

Example 5.68

```
Employee_Record = [['501', 'R.Naveen Kumar', '9876543210',
130000],
        ['502', 'J.Srikanth', '9988334477', 125000],
        ['503', 'Mahipal Reddy', '8899223377', 60000]]
print(Employee_Record[2][1][4])
```

Employee_Record[2][1][4] means, in the second list, find the first item and its fourth character. Here the second list is ['503', 'Mahipal Reddy', '8899223377', 60000], the first item is 'Mahipal Reddy' and its fourth character is 'p'. Remember that the count starts with 0.

Output

```
p
```

Example 5.69

```
Employee_Record = [['501', 'R.Naveen Kumar', '9876543210',
130000],
        ['502', 'J.Srikanth', '9988334477', 125000],
        ['503', 'Mahipal Reddy', '8899223377', 60000]]
print(Employee_Record[2][3][4])
```

Here the fourth character of the third item of the second list is not a String datatype. So we get an error message.

Output

```
Traceback (most recent call last):
  File "C:\Users\RakeshN\AppData\Local\Programs\Python\Py
  thon38-32\chap-5.py", line 16, in <module>
    print(Employee_Record[2][3][4])
TypeError: 'int' object is not subscriptable
```

Example 5.70 Row sum of a matrix

```
Mat = [[501, 453, 664],
         [502, 444, 646],
         [503, 222, 999]]
for i in range (len(Mat)):
  sum = 0
  for j in range(len(Mat[i])):
    sum = sum + Mat[i][j]
print("The sum of ", Mat[i],"is : ", sum)
```

Output

```
The sum of [501, 453, 664] is : 1618
The sum of [502, 444, 646] is : 1592
The sum of [503, 222, 999] is : 1724
```

Example 5.71

```
Mat1 = [[53, 45, 66],
    [52, 44, 63],
    [50, 22, 99]]
Mat2 = [[1, 2, 3],
    [4, 5, 6],
    [7, 8, 9]]
Mat3 = [[0, 0, 0],
    [0, 0, 0],
    [0, 0, 0]]
print("The sum of two matrix is :")

for i in range (len(Mat1)):
  sum = 0
  for j in range(len(Mat1[i])):
    Mat3[i][j] = Mat1[i][j] + Mat2[i][j]
  print(Mat3[i])
```

Output

```
The sum of two matrix is :
[54, 47, 69]
[56, 49, 69]
[57, 30, 108]
```

Example 5.72 Matrix multiplication

```
m = int(input("How many rows in Matrix1? "))
n = int(input("How many columns in Matrix1? "))
Mat1 = [[0 for x in range(n)] for x in range (m)]
  for i in range(m):
    for j in range(n):
      Mat1[i][j] = int(input("Enter the elements of Matrix 1: "))

m1 = int(input("How many rows in Matrix2? "))
n1 = int(input("How many columns in Matrix2? "))

if (n != m1):
  print("Matrix multiplication NOT possible!!!")
  exit()

Mat2 = [[0 for x in range(n1)] for x in range (m1)]
for i in range(m1):
  for j in range(n1):
    Mat2[i][j] = int(input("Enter the elements of Matrix 2: "))
```

```
print("\nMatrix 1 is ")
for i in range(m):
  for j in range(n):
    print(Mat1[i][j], end = " ")
  print()
print("\nMatrix 2 is ")
for i in range(m1):
  for j in range(n1):
    print(Mat2[i][j], end = " ")
  print()

Mat3 = [[0 for x in range(m)] for x in range (n1)]

for i in range(m):
  for j in range(n1):
    for k in range(m1):
      Mat3[i][j] += Mat1[i][k] * Mat2[k][j]

print("\nMatrix multiplication is ")
for i in range(m):
for j in range(n1):
print(Mat3[i][j], end = " ")
print()
```

Output

```
How many rows in Matrix1? 2
How many columns in Matrix1? 3
Enter the elements of Matrix 1: 1
Enter the elements of Matrix 1: 2
Enter the elements of Matrix 1: 3
Enter the elements of Matrix 1: 4
Enter the elements of Matrix 1: 5
Enter the elements of Matrix 1: 6
How many rows in Matrix2? 3
How many columns in Matrix2? 2
Enter the elements of Matrix 2: 6
Enter the elements of Matrix 2: 8
Enter the elements of Matrix 2: 9
Enter the elements of Matrix 2: 0
Enter the elements of Matrix 2: 6
Enter the elements of Matrix 2: 4

Matrix 1 is
1 2 3
4 5 6

Matrix 2 is
6 8
9 0
6 4
```

```
Matrix multiplication is
42 20
105 56
```

5.4   TUPLES

A Python tuple is an immutable collection of objects separated by a comma.
The general syntax for a tuple is:

Variable_Name = (element1, element2, element3, ..., element n)
   or
Variable_Name = element1, element2, element3, ..., element n

Where elements can be any Python object. In a tuple notice that:

- The objects are separated by a comma.
- A pair of parentheses enclosing the objects is not compulsory except while creating an empty tuple. A pair of parenthesis improves readability.
- The objects in a tuple are immutable.
- As the objects inside the tuple are immutable, so a tuple is always initialized.

Example 5.70

```
Tup1 = ('Crow', 'Peacock', 'owl', 'Parrot')
print(Tup1)
print(type(Tup1))
Tup2 = 'Crow', 'Peacock', 'owl', 'Parrot'
print(Tup2)
print(type(Tup2))
```

Observe that *Tup1* is a tuple which is a collection of a few bird names separated by a comma inside a pair of parentheses and *Tup2* is the same collection of bird names separated by a comma but without any parenthesis. It can be observed that in either way the type is *'tuple'*. So, the pair of parentheses is not mandatory for a tuple.

Output

```
('Crow', 'Peacock', 'owl', 'Parrot')
<class 'tuple'>
('Crow', 'Peacock', 'owl', 'Parrot')
<class 'tuple'>
```

5.4.1   CREATING A TUPLE

5.4.1.1   Creating a Tuple with no Item

A tuple with no item can be created simply by a pair of parentheses with nothing inside it.

Example 5.71

```
tup = ()
print("The tuple has elements ", tup)
print(type(tup))
```

Observe that by simply not keeping any element inside the parenthesis we can create a blank tuple.

Output

```
The tuple has elements ()
<class 'tuple'>
```

5.4.1.2   Creating a Tuple with a Single Item

A tuple with a single item cannot be created simply by putting an element inside a parenthesis. This way it will create a String but not a tuple.

Example 5.73

```
tup1 = ('Peacock')
print('Type of tup1 is ', type(tup1))
tup2 = 'Peacock'
print("Type of tup2 is ", type(tup2))
```

Observe that by simply keeping an element inside a pair of parentheses or without any parenthesis creates a String but not a tuple.

Output

```
Type of tup1 is <class 'str'>
Type of tup2 is <class 'str'>
```

To create a tuple with a single item, a comma is used after the element.

Example 5.74

```
tup1= ('Peacock',)
print("Type of tup1 is ", type(tup1))
tup2 = 'Peacock',
print("Type of tup2 is ", type(tup2))
```

Observe that after the element *'Peacock'* a comma is used with/without a pair of parentheses. This created a tuple with a single item.

Output

```
Type of tup1 is <class 'tuple'>
Type of tup2 is <class 'tuple'>
```

5.4.1.3 Nesting Tuples

Tuple can be defined as a collection of objects, other tuples and lists.

Example 5.74

```
lst1 = ['Crow','Peacock']
tup2 = ('Parrot', 'Sparrow')
tup1 = ('owl', lst1, tup2)
print("New Tuple is ", tup1)
```

Observe that *lst1* is a list and *tup2* is a tuple. *Tup1* is one more tuple defined by taking an element *'owl'*, *lst1* and *tup2*.

Output

```
New Tuple is ('owl', ['Crow', 'Peacock'], ('Parrot',
'Sparrow'))
```

5.4.1.4 Creating a Tuple from List

A list can also be converted into a tuple. This can be done by type casting a *tuple*.

Example 5.75

```
lst1 = ['Crow', 'Peacock']
print("The list is ", lst1)
tup = tuple(lst1)
print("New Tuple is ", tup)
```

Observe the third line. The list, *lst1*, is typecast and converted to a tuple.

Output

```
The list is ['Crow', 'Peacock']
New Tuple is ('Crow', 'Peacock')
```

5.4.1.5 Creating a Tuple from a String

A String can also be converted into a tuple. This can be done by type casting a *tuple*.

Example 5.75

```
str1 = "Python"
print("The given String is: ", str1)
tup = tuple(str1)
print("The tuple is: ", tup)
```

Observe the third line. The String, *str1*, is typecast and converted to a tuple where each alphabet of the String is converted as an item in the tuple.

Output

```
The given String is: Python
The tuple is: ('P', 'y', 't', 'h', 'o', 'n')
```

5.4.2 TUPLE OPERATIONS

5.4.2.1 Concatenation (+) Operation

A single tuple can be created by concatenating more than one tuple. Concatenation is done by using the "+" symbol.

Example 5.76

```
Tup1 = ('Peacock', 'owl', 'Crow')
Tup2 = ('Parrot', 'Sparrow', 'Crow')
print("The first Tuple is: ", Tup1)
print("The second Tuple is: ", Tup2)

Tup = Tup1 + Tup2
print("The new Tuple is: ", Tup)
```

Output

```
The first Tuple is: ('Peacock', 'owl', 'Crow')
The second Tuple is: ('Parrot', 'Sparrow', 'Crow')
The new Tuple is: ('Peacock', 'owl', 'Crow', 'Parrot',
'Sparrow', 'Crow')
```

5.4.2.2 Repetition (*) Operation

A single tuple can be created with repeated objects in it. Repetitions are done by using "*" symbol. The number of times to be repeated must be an Integer.

Example 5.77

```
Tup1 = ('Peacock', 'owl', 'Crow')
print("The first Tuple is ", Tup1)
Tup = Tup1 * 2
print("The Tuple with repeated item is \n", Tup)
```

Observe the third line *Tup=Tup1*2*. This caused all the *objects of Tup1* to be repeated twice in tuple *Tup*.

Output

```
The first Tuple is ('Peacock', 'owl', 'Crow')
The Tuple with the repeated item is
('Peacock', 'owl', 'Crow', 'Peacock', 'owl', 'Crow')
```

5.4.3 TUPLE ASSIGNMENT

Tuple has a very strong assignment feature. We have seen in the definition of tuple, that the objects on the right-hand side may be inside a pair of parentheses, separated by a comma assigned to a variable, which is on the left-hand side. This is called tuple packing.

Example 5.78 Tuple packing

```
Tup1 = ('Crow', 'Peacock', 'owl', 'Parrot')
print(Tup1)
```

In tuple packing, we assign the data objects to a variable. It becomes easy to access all the data objects at a time, just by using the tuple name.

Output

```
('Crow', 'Peacock', 'owl', 'Parrot')
```

In tuple unpacking, the data objects are in the tuple and are assigned to separate variables on the left-hand side.

Example 5.79 Tuple unpacking

```
Tup1 = ('Crow', 'Peacock', 'Owl', "parrot", 'Sparrow')
Bird1, Bird2, Bird3, Bird4, Bird5 = Tup1
print("First bird is ", Bird1)
```

Here observe that each bird now can be accessed with the help of a variable name. When there are few data objects in the Tuple, it is easy to access each element with a separate variable name.

Output

```
First bird is Crow
```

While Tuple Unpacking, one thing to remember is, we must assign a variable name to each element of the tuple, otherwise it will raise a *ValueError*.

Example 5.80 Tuple unpacking

```
Tup1 = ('Crow', 'Peacock', 'Owl', 'parrot', 'Sparrow')
Bird1, Bird2, Bird3 = Tup1
```

Observe that there are five bird names in the tuple *tup1*. While unpacking, only three variables are used so a value error occurred.

Output

```
Traceback (most recent call last):
```

```
File "C:\Users\RakeshN\AppData\Local\Programs\Python\Py
thon37-32\chap-5.py", line 125, in <module>
Bird1, Bird2, Bird3 = Tup1
```

ValueError: too many values to unpack (expected 3)

It is also possible to interchange the tuples like any other variable.

Example 5.81 Tuple exchange

```
Tup1 = ('Crow', 'Peacock', 'Owl', 'parrot', 'Sparrow')
Tup2 = ('Mango', 'Apple', 'Banana', 'Orange', 'Grape')
print("First tuple: ", Tup1)
print("Second tuple: ", Tup2)
(Tup1, Tup2) = (Tup2, Tup1)
print("\n First tuple: ", Tup1)
print(" Second tuple: ", Tup2)
```

Output

```
First tuple: ('Crow', 'Peacock', 'Owl', 'parrot', 'Sparrow')
Second tuple: ('Mango', 'Apple', 'Banana', 'Orange',
'Grape')

First tuple: ('Mango', 'Apple', 'Banana', 'Orange', 'Grape')
Second tuple: ('Crow', 'Peacock', 'Owl', 'parrot',
'Sparrow')
```

5.4.4 ACCESSING OBJECTS IN TUPLE

Just like a list, objects in a tuple can also be accessed with the help of an index. Negative indices are legal in python. An element with an index equal to –1 is the last element in the tuple. Accessing the tuple objects from the left-to-right start with index 0 and keeps on incrementing by 1 until the end of the tuple, whereas accessing the tuple objects from the right-to-left starts with index –1 and keeps decrementing by 1 until the beginning of the tuple. This can be seen in figure 5.1. in section 5.2.3.

5.4.4.1 Membership in Tuple

The keyword *in* is a membership function. It checks if an element is in the tuple or not. It returns a Boolean value. If the given element is in the tuple, it returns True otherwise it returns False.

Example 5.83

```
Tup1 = ('Crow', 'Peacock', 'owl', 'Parrot', 'Sparrow')
x = 'Peacok'
```

```
if x in Tup1:
  print(x," exists in the Tuple.")
else:
  print(x," is not in the Tuple.")
```

Output

```
Peacok is not in the Tuple.
```

5.5.5 UPDATING TUPLES

As per the definition, a tuple is an immutable collection of objects. It means once the objects in the tuple are defined, it cannot be changed. An effort to change the element results in an error.

Example 5.84

```
Tup1 = ('Crow', 'Peacock', 'owl', 'Parrot', 'Sparrow')
print("Items in the tuple index ", Tup1)
Tup[0] = 'Cock'
print("Items in the tuple index ", Tup2)
```

Observe that changing the first element of the tuple results in TypeError.

Output

```
Items in the tuple index ('Crow', 'Peacock', 'owl',
'Parrot', 'Sparrow')
Traceback (most recent call last):
  File "C:\Users\RakeshN\AppData\Local\Programs\Python\Py
  thon37-32\chap-5.py", line 183, in <module>
  Tup[0] = 'Cock'
NameError: name 'Tup' is not defined
```

However, if a tuple contains a list, the item inside the list can only be updated, inserted or deleted.

Example 5.85

```
Tup1 = ('Crow', 'Peacock', 'owl', ['Parrot', 'Sparrow'])
print("The Original tuple is ", Tup1)
Tup1[3][0]=45
print("The tuple after updation ", Tup1)
Tup1[3].insert(0, 'Swan')
print("The tuple after insertion ", Tup1)
del Tup1[3][1]
print("The tuple after deletion ", Tup1)
```

Observe that in the tuple, *Tup1*. It contains a list with two objects 'Parrot' and 'Sparrow'. While traveling from left-to-right, 'Crow' is the 0th index and the list is the 3rd index. The objects inside the list can be accessed by one more index. It means *Tup1[3][0]* has value 'Parrot' and *Tup1[3][1]* contains 'Sparrow'.

In the third line of the program, the value at *Tup1[3][0]* is changed to 45. In the fifth line of the program, a new value 'Swan' is inserted at *Tup1[3][0]* location. In the seventh line, the element at location *Tup1[3][1]* is deleted, which is reflected in the output.

Output

```
The Original tuple is ('Crow', 'Peacock', 'owl', ['Parrot',
'Sparrow'])
The tuple after updation ('Crow', 'Peacock', 'owl', [45,
'Sparrow'])
The tuple after insertion ('Crow', 'Peacock', 'owl',
['Swan', 45, 'Sparrow'])
The tuple after deletion ('Crow', 'Peacock', 'owl', ['Swan',
'Sparrow'])
```

5.4.6 TUPLE FUNCTIONS AND METHODS

Just like functions in String and list, we have functions in tuple. The syntax and use of the functions in the tuple are the same as that of String and list. The functions like *len()*, *max()* and *min()* behave exactly the same as a list. It is advisable for the reader to refer to section 5.2.4. The methods available in the tuple are also the same as that of the list such as *Count*, *Index*, etc**.**

The function like *del()* is not applicable as the tuple is immutable.

5.4.7 TUPLE SLICING

Slicing in a tuple is the same as in a list. Slicing is a flexible tool to build a new tuple out of an existing tuple. Python supports slice notation for any sequential datatype like lists, strings, tuples, bytes, byte-arrays and ranges. Slicing allows extracting or referring to a section of the tuple, like a slice in a cake. The general syntax is: *tuple_Name=List[start : end : step]*

Where the *start* is the starting index, the *end* is the ending index and *step* (optional) is the Integer to skip before getting the next element. Remember that the start index is always less than the *end* index, otherwise, it returns an empty tuple. And the element in the end index is not included. It is advisable to refer to section **5.2.4.**

5.5 ADVANTAGES OF TUPLE OVER LIST

Since tuples are quite similar to lists, both of them are used in similar situations as well.

However, there are certain advantages of implementing a tuple over a list. Listed below are some of the main advantages:

1. Since tuples are immutable, iterating through tuples is faster than with a list. So there is a slight performance boost.
2. Tuples that contain immutable objects can be used as a key for a dictionary. With lists, this is not possible.
3. If we have data that doesn't change, implementing it as a tuple will guarantee that it remains write-protected.

Example 5.85 Write a program that creates a tuple by reading objects supplied by a user

```
no = int(input("Enter how many elements in the Tuple: "))
Tup = ()
for i in range(no):
  element = eval(input("Enter elements for the Tuple: "))
  Tup = Tup + (element,)
print("The Tuple is : ",Tup)
```

Output

```
Enter how many elements in the Tuple: 5
Enter element for the Tuple: 1
Enter element for the Tuple: 'Parrot'
Enter element for the Tuple: ['Mango','Apple']
Enter element for the Tuple: 5.5
Enter element for the Tuple: 'Tiger',56
The Tuple is : (1, 'Parrot', ['Mango', 'Apple'], 5.5,
('Tiger', 56))
```

Example 5.85 Write a program that creates a list of tuples where each tuple contains a pair, the element and its square in a range supplied by a user. Example: input: 3; output: [(0,0),(1,1),(2,4)].

```
no = int(input("Enter how many elements in the Tuple: "))
Tup = tuple[(x, x**2) for x in range(no)]
print(Tup)
```

Output

```
Enter how many elements in the Tuple: 3
((0, 0), (1, 1), (2, 4))
```

Example 5.86 Write a program that creates a String from the objects of a tuple. Example: input: ('P','y','t','h','o','n'); output: Python.

```
Tup1 = ('P', 'y', 't','h','o','n')
print("The tuple is ", Tup1)
str1= ''.join(Tup1)
print("The String is ", str1)
```

Output

```
The tuple is ('P', 'y', 't', 'h', 'o', 'n')
The String is Python
```

Example 5.86 Write a program that reads two indexes in a tuple. The first index is less than the second index otherwise an appropriate message is to be displayed. Find the sum of digits between these two indexes (objects at both indexes included).

```
Tup1 = (1,4,2,6,4,7,6,4,68,96,4)
print("The given tuple is ", Tup1)
no1 = int(input("Enter the first index of "))
no2 = int(input("Enter the first index of "))
if((no1 >= no2) or (no1 >= len(Tup1)) or (no2 >
len(Tup1))):
  print("Try again ....")
else:
  Tup2 = Tup1[no1:no2+1]
  print("The tuple between", no1, " index and ", no2, "
  index is ", Tup2)
  sum = 0
  for i in Tup2:
    sum +=i
print("The sum of digits between", no1, " index and ",
no2, " index is ", sum)
```

Output

```
The given tuple is (1, 4, 2, 6, 4, 7, 6, 4, 68, 96, 4)
Enter the first index of 2
Enter the first index of 8
The tuple between index 2 and index 8 is (2, 6, 4, 7, 6, 4, 68)
The sum of digits between index 2 and index 8 is 97
```

MULTIPLE CHOICE QUESTIONS – LIST

1. Find the output of the following code
   lst = [1, 2, 3, 4, 5]
   print(lst[−1])
   a. 2
   b. 3
   c. 4
   d. 5

2. Find the output of the following code
   lst = [1, 2, 3, 4, 5]
   print(list(lst[–3:–1]))

   a. [2, 3]
   b. [3, 4]
   c. [4, 5]
   d. [3, 5]

3. Find the output of the following code
   lst = [0, 1, 2, 3, 4, 5]
   lst.insert(0,1)
   del lst[1]
   print(lst)

   a. [0, 1, 2, 3, 4, 5]
   b. [1, 1, 2, 3, 4, 5]
   c. [2, 1, 2, 3, 4, 5]
   d. [3, 1, 2, 3, 4, 5]

4. Find the output of the following code
   lst = [0, 1, 2, 3, 4, 5]
   lst1 = lst
   del lst1[1:2]
   print(lst1)

   a. [0, 1, 3, 4, 5]
   b. [1, 2, 3, 4, 5]
   c. [0, 2, 3, 4, 5]
   d. [0,1, 3, 4, 5]

5. Find the output of the following code
   lst = [0, 1, 2, 3, 4, 5]
   lst1 = lst
   del lst1[-1:-2]
   print(lst1)

   a. lst and lst1 have same length
   b. raise an error
   c. lst1 is a blank list
   d. lst1 is longer than lst

6. Find the output of the following code
   lst = [0, 1, 2,3, 4, 5]
   lst1 = []
   for i in lst:
        lst1.insert(0,i)
   print(lst1)

   a. [0, 1, 2, 3, 4, 5]
   b. [5, 4, 3, 2, 1, 0]
   c. [0, 0, 0, 0, 0]
   d. None of the above

7. Find the output of the following code
```
lst = [0, 1, 2, 3, 4, 5]
for i in range(len(lst)):
    lst.insert(-1,lst[i])
print(lst)
```

   a. [1, 2, 3, 4, 5, 0, 0, 0, 0, 0, 0, 0]
   b. [1, 0, 0, 0, 0, 0, 0, 1, 2, 3, 4, 5]
   c. [0, 0, 0, 0, 0, 0, 0, 1, 2, 3, 4, 5]
   d. [0, 1, 2, 3, 4, 0, 1, 2, 3, 4, 0, 5]

8. Find the output of the following code
```
lst = [[0, 1, 2, 3, 4, 5] for i in range(3)]
print(lst[2][1])
```

   a. 0
   b. 1
   c. 2
   d. 3

9. Find the output of the following code
```
my_list = [1, 2, 3, 4, 5, 6]
count = 0
for item in my_list:
    if type(item) == str:
        contunue
    count = count + 1
print(count)
```

   a. 2
   b. 4
   c. 6
   d. 0

10. Find the output of the following code
```
m = 0
my_list_1= [1, 2, 5]
my_list_2= [1, 3, 2, 6, 5]
for x in my_list_1:
    for y in my_list_2:
        if x == y:
            m = m+1
print(m)
```

   a. 3
   b. 4
   c. 6
   d. 0

11. Find the output of the following code

```
n1 = [10, 20, 30, 40, 50]
n2 = [10, 20, 30, 40, 50]
print(n1 is n2)
print(n1 == n2)
n1 = n2
print(n1 is n2)
print(n1 == n2)
```

a. True, True, True, False
b. True, False, True, True
c. True, True, False, True
d. False, True, True, True

12. Find the output of the following code

```
prices = [30.5, '40.5', 10.5]
total = 0
for price in prices:
total += price
print(total)
```

a. 81.5
b. 30.1, 71.0, 81.5
c. Expected Indentation
d. Raise an error ValueError

13. Find the output of the following code

```
str1 = "I am learing Python"
str2 = str1.split('a')
print(str2)
```

a. ['I ', 'm le', 'rning Python']
b. ['Ia ', 'm lea', 'rning Python']
c. ['I ', 'am le', 'arning Python']
d. ['I ', 'am', 'learning', 'Python']

14. Find the output of the following code

```
numbers = [0, 1, 2, 3, 4, 5, 6, 7, 8, 9]
index = 0
while index < 10:
    print(numbers[index])
    if numbers(index) == 6:
        break
    else:
        index += 1
print(index)
```

a. 0, 1, 2, 3, 4, 5
b. 1, 2, 3, 4, 5, 6
c. 0, 1, 2, 3, 4, 5, 6
d. Raise an Error

15. Find the output of the following code
```
fruits = ['apple', 'orange', 'mango', 'banana']
for i in range(len(fruits)):
    fruits[i] = fruits[i][-1].upper()
print(fruits)
```

    a.   ['B', 'A', 'N', 'A', 'N', 'A']
    b.   ['A', 'O', 'M', 'B']
    c.   ['E', 'E', 'O', 'A']
    d.   ['L', 'G', 'G', 'N']

16. Find the output of the following code
```
i = [10, 20,[30, 40],[50, 60]]
count = 0
for i in range(len(l)):
    if type(l[i]) == list:
        count = count + 1
print(count)
```

    a.   1
    b.   2
    c.   3
    d.   4

17. Find the output of the following code
```
x = [13, 4, 17, 10]
w = x[1:]
u = x[1:]
y = x
u[0] = 50
y[1] = 40
print(x)
```

    a.   [13, 40, 17, 10]
    b.   [50, 40, 10]
    c.   [13, 4, 17, 10]
    d.   [50, 40, 17, 10]

18. Find the output of the following code
```
a = ['a', 'b', 'c', 'd']
for i in a:
    a.append(i.upper())
print(a)
```

    a.   ['a', 'b', 'c', 'd']
    b.   ['A', 'B', 'C', 'D']
    c.   Raise an Error
    d.   None of the above

19. Find the output of the following code
```
numbers = [0, 1, 2, 3, 4, 5, 6, 7, 8, 9]
i = '0'
index = 0
while(index < 10):
    if numbers[index] == 4:
        break
    else:
        index += 1
        i = i + str(index)
print(i)
```

a. 1234
b. 0123
c. 12345
d. 01234

20. Find the output of the following code
```
my_list = [1, 2, 3, 4, 5, 6]
count = 0
for item in my_list:
    if type(item != str):
        continue
    count += 1
print(count)
```

a. 3
b. 2
c. 1
d. 0

MULTIPLE CHOICE QUESTIONS – TUPLE

1. The += operator when applied to a tuple, performs
   a. Concatenation
   b. Subtraction
   c. Multiplication
   d. All of the above

2. Find the output of the following code
```
n1 = (1, 2, 3, 4, 5)
n2 = (1, 2, 3, 4, 5)
print(n1 is n2)
print(n1 == n2)
```

```
n1 = n2
print(n1 is n2)
print(n1 == n2)
```

   a. False, True, True, True
   b. True, True, False, True
   c. True, False, True, True
   d. True, True, True, True

3. Which among the following is correct for A=('Mango',2,'Three',4.0,'Four',5)
   a. A[1]='Mango'
   b. A[2]=A[-4]
   c. A[3]='Three'
   d. A[3]=A[-3]

4. Find the output of the following code

```
a = ()
print(bool(a))
```

   a. False
   b. True
   c. Depends on Processor
   d. Not Defined

5. Find the output of the following code

```
prices = (30.5, '40.5', 10.5)
    total = 0
for price in prices:
total += price
print(total)
```

   a. AttributeError
   b. TypeError
   c. ValueError
   d. None of the above

6. In the following code

```
prices = (30.5, '40.5', 10.5)
total = 0
for price in prices:
    total += # write your code here
print(total)
```

Which option to be substituted in place of comment to get an output
   a. tuple(price)
   b. int(price)
   c. float(price)
   d. str(price)

7. In the following code
   numbers = (10, 20, 30, 40)
   x = 0

   Which option will get an output 10
   a. for i in (30, 40, 50):
            if i not in numbers:
                x = x + 5
          print(x)
   b. for i in (30, 40, 50):
            if i not in numbers:
                x = x + 10
          print(x)
   c. for i in (30, 40, 50):
            if i in numbers:
                x = x + 5
          print(x)
   d. for i in (30, 40, 50):
            if i not in numbers:
                x = x + 10
          print(x)

8. Find the output of the following code
   numbers = (10, 20, 30, 40, 50)
   alphabets = ('a', 'b', 'c', 'd', 'e')
   print(numbers is alphabets)
   print(numbers == alphabets)
   numbers = alphabets
   print(numbers is alphabets)
   print(numbers == alphabets)

   a. False, False, False, False
   b. False, True, True, False
   c. False, True, False, True
   d. False, False, True, True

9. In the following code
   sub1 = ('java','python','C')
   sub2 = ('java','python','C')
   sub3 = sub1
   print(sub1 is sub2)

   Which option will print True.
   a. print(sub1 == sub2)
   b. print(sub1 is sub2)
   c. print(sub3 is sub2)
   d. All of the above

10. Find the output of the following code

```
tup1 = ('java', 'python', 'C')
tup2 = 'C#'
print(tup1)
```

 a. ('java', 'python', 'C#')
 b. 'java', 'python', 'C'
 c. 'java', 'python', 'C#'
 d. Error

11. Find the output of the following code

```
tup1 = (['java','python'],'C')
tup1[0][0] = 'R'
print(tup1)
```

 a. (['R', 'python'], 'C')
 b. (['python', 'R'], 'C')
 c. ('R', 'python', 'C')
 d. Error

12. Find the output of the following code

```
values = ([3,4,5,1],[33,6,1,2])
v = values[0][0]
for lst in values:
    for element in lst:
        if v > element:
            v = element
print(v)
```

 a. 1
 b. 2
 c. 3
 d. 33

13. Fin d the output of the following code

```
values = (3, 4, 5, 1, 33, 6, 1, 2)
val = values[::-1]
print(val)
```

 a. (3, 4, 5, 1, 33, 6, 1, 2)
 b. (2, 1, 6, 33, 1, 5, 4, 3)
 c. (4, 1, 6, 2)
 d. (3, 5, 33, 1)

14. Fin d the output of the following code

```
values = (3, 4)
a,b = values
```

```
a,b = b,a
print(values)
```

a. (3, 4)
b. (4, 3)
c. (3, 3)
d. (4, 4)

15. Fin d the output of the following code

```
a = ('a', 'b', 'c', 'd')
for i in a:
a.append(i.upper())
print(a)
```

a. ('a', 'b', 'c', 'd', 'A', 'B', 'C', 'D')
b. ('A', 'B', 'C', 'D', 'a', 'b', 'c', 'd')
c. ValueError
d. AttributeError

16. Fin d the output of the following code

```
lst = (7,8,9)
b = lst[:]
print(b is lst)
print(b == lst)
```

a. False, False
b. False, True
c. True, False
d. True, True

17. Find the output of the following code

```
tup = (1, 2, 3, 4, 5)
count = 0
for i in tup:
    if (type(i) == float):
        count += 1
print(count)
```

a. 0
b. 1
c. 2
d. 3

18. Find the output of the following code

```
tup = (1, 2, 3, 4, 5)
sum = 0
for i in tup:
    if (type(i) == float):
```

```
        pass
    else:
        sum += float(i)
print(sum)
```

a. 21
b. 12.0
c. 21.0
d. 12

19. Which of the following in tup=(0, 1, 2, 3, 4) is incorrect
    a. print(tup[3])
    b. print(len(tup))
    c. print(max(tup))
    d. tup[3]=12

20. Find the output of the following code
```
tup = (1, 2, 4, 3, 8, 9)
x = [tup[i] for i in range(0,len(tup),2)]
print(x)
```

a. [1, 2, 4, 3, 8, 9]
b. [2, 3, 9]
c. [1, 4, 8]
d. (1, 4, 8)

DESCRIPTIVE QUESTIONS – LIST

1. What are the different sequential datatypes? Explain what the different ways of creating a list are.
2. Describe the negative index for a sequence.
3. Explain operations in a list.
4. Explain default parameters in slicing
5. Explain the difference between the assignment of a list and copying of a list.

DESCRIPTIVE QUESTIONS – TUPLE

1. What are the differences between a tuple and a list?
2. What is packing and unpacking of a tuple?
3. What are the different operations in a tuple?
4. How do you create a tuple with a single element?
5. How do you create a tuple from a String and vice-versa?
6. Is it possible to make any changes to the objects in a tuple? Explain your answer with an example.

PROGRAMMING QUESTIONS – LIST

1. Write a Python program to remove duplicates from a list.
2. Write a Python program to find the list of words that are longer than n from a given list of words.

3. Write a Python program to replace the last element in a list with another list.
4. Write a Python program to insert a given String at the beginning of all items in a list. Example Input list : [1, 2, 3, 4], RollNo output: ['RollNo1', 'RollNo2', 'RollNo3', 'RollNo4']
5. Write a Python program to iterate over two lists simultaneously. Example input list1: [1, 2, 3, 4], list2:['cat', 'rat', 'dog', 'pet']
   output:
   1. cat
   2. rat
   3. dog
   4. pet
6. Write a Python program to find the list in a list of lists whose sum of elements is the highest. Example input list: [[1, 2, 3],[3, 6, 3],[65, 3, 754]] output: 822
7. Write a Python program that will add square brackets to a list. If the list contains a single item ['Peacock'] and we want 3 more square brackets [ ], the output should be [[[['Peacock']]]]

PROGRAMMING QUESTIONS – TUPLE

1. Write a Python program to remove the duplicate objects from a tuple.
2. Write a program that removes an item at a specific location.
3. Write a Python program to find the repeated objects of a tuple.
4. Write a Python program to replace the last value of tuples in a tuple with a specific value.
5. Write a Python program to remove an empty tuple(s) from a list of tuples.
6. Write a Python program to sort a tuple of tuples by its second item as a String element.

ANSWER TO MULTIPLE CHOICE QUESTIONS – LIST

1.	D	2.	B	3.	B	4.	C	5.	C
6.	B	7.	D	8.	B	9.	C	10.	A
11.	D	12.	C	13.	A	14.	D	15.	C
16.	B	17.	A	18.	C	19.	D	20.	D

ANSWER TO MULTIPLE CHOICE QUESTIONS – TUPLE

1.	A	2.	D	3.	B	4.	A	5.	B
6.	C	7.	A,C	8.	D	9.	D	10.	D
11.	D	12.	A	13.	B	14.	A	15.	D
16.	D	17.	A	18.	C	19.	D	20.	C

6 Dictionary

LEARNING OBJECTIVES

After studying this chapter, the reader will be able to:

- Create/initialize a dictionary
- Access elements in a dictionary
- Perform different operations in a dictionary
- Use some common dictionary functions and methods

6.1 INTRODUCTION

Dictionary is a non-sequential and the only mapping type data structure. Mapping refers to a one-to-one correspondence between the key and its value. It provides a fast lookup by keys. We can think of a to-do list. Corresponding to every number in a to-do list, there is a description. The numbers in a to-do list are unique, where the description may be the same.

6.2 DICTIONARIES

A dictionary in Python is an unordered collection of items. Each item in a dictionary has a key, and each key is associated with a value. Items in a dictionary do not have any order. Dictionary is treated as a bag of items rather than a sequence of items. Since items in a dictionary are not in sequence, we cannot use an index to refer to an item in a dictionary. The general syntax is:

Dictionary_Name = {key 1:value 1, key 2:value 2, ..., key n: value n}

In a dictionary notice that:

- A dictionary *starts* and *ends* with a pair of *curly brackets { }*.
- The elements are filled within the curly brackets in the form of a key and a value pair separated by a *colon*. Each key-value pair is separated by a *comma*.
- The elements inside a dictionary may have *different types*.
- The keys and the values in a dictionary may be *numeric* or *String*.
- Each key must be *unique*.
- A dictionary is a *one-way* tool. It means, if a key is given, the corresponding value can be retrieved, but not vice versa.
- Refer to an item in a dictionary by its key.

DOI: 10.1201/9781003219125-7

Example 6.1 Sample dictionary

```
Bird_sound = {'crow':'caw','Doves':'coo','Ducks':'quack',
'Owl': 'hoot'}
Employee = {'Name':'Ram', 'Age':30, 'Salary':12500.50}
Animals = {1:'cat', 2:'cow', 3:'camel'}
Phone = {1:9400530306, 2:90305404093, 3:8247891965}
Skills = {1 : ['C','java'],2:('python','C#')}
people = {('Hari','Sita'):55}
print(Bird_sound)
print(Employee)
print(Animals)
print(Phone)
print(Skills)
print(people)
```

Observe that the keys and the values can be any combination of numeric, String, list and tuple values except list as a value of the dictionary.

Output

```
{'crow': 'caw', 'Doves': 'coo', 'Ducks': 'quack', 'Owl':
'hoot'}
{'Name': 'Ram', 'Age': 30, 'Salary': 12500.5}
{1: 'cat', 2: 'cow', 3: 'camel'}
{1: 9400530306, 2: 90305404093, 3: 8247891965}
{1: ['C', 'java'], 2: ('python', 'C#')}
{('Hari', 'Sita'): 55}
```

6.2.1   CREATING A DICTIONARY

6.2.1.1   Creating a Dictionary with no item

A dictionary with no item can be created simply by a pair of curly brackets or using the keyword *dict()* followed by a pair of parentheses with nothing inside it.

Example 6.2

```
my_dict1= {}
my_dict2= dict()
print("The dictionary are ", my_dict1, " and ", my_dict2)
print("The type is ", type(my_dict1), " and ",
type(my_dict2))
```

Output

```
The dictionary are {} and {}
The type is <class 'dict'> and <class 'dict'>
```

6.2.1.2 Creating a Dictionary with { }

In general a dictionary is created as described in example 6.1.

Example 6.3

```
my_dict1 = {1:'cat', 2:'cow', 3:'camel', 4:'cheetah'}
print('The dictionary is ', my_dict1)
```

6.2.1.3 Creating a Dictionary with dict( )

A dictionary can also be created by the keyword *dict()*. The parameter passed to this function is a pair of curly brackets and the elements are filled within the curly brackets in the form of a key and a value pair separated by a colon. Each key-value pair is separated by a comma.

Example 6.4

```
my_dict1 = dict({1:'cat', 2:'cow', 3:'camel', 4:'cheetah'})
print("The dictionary is ", my_dict1)
```

The parameter passed to this function is a list and the elements are filled within the square brackets in the form of a tuple, where each key and value pair is separated by a comma. Each tuple is separated by a comma.

Example 6.5

```
my_dict1 = dict([ [1,'cat'],[2,'cow'],
[3,'camel'],[4,'cheetah'] ])
print("The dictionary is ", my_dict1)
```

The parameter passed to this function is a tuple and the elements are filled within the square brackets in the form of a tuple, where each key and value pair is separated by a comma. Each list is separated by a comma.

Example 6.6

```
my_dict1 = dict(([1,'cat'],[2,'cow'], [3,'camel'],[4,'cheetah']))
print("The dictionary is ", my_dict1)
```

Output

```
The dictionary is {1: 'cat', 2: 'cow', 3: 'camel', 4:
'cheetah'}
```

6.2.1.4 Creating a Default Dictionary

A default directory is a directory in which for every key, the value that remained the same may also be *None*. To create a default dictionary *formkeys()* is used.

Example 6.7

```
my_dict1= {}.fromkeys((1,2,3,4),'cat')
print("The Dictionary is ", my_dict1)
```

Output

```
The Dictionary is {1: 'cat', 2: 'cat', 3: 'cat', 4: 'cat'}
```

6.2.1.5 Creating a Dictionary from a List

A dictionary can be created by combining two lists of the same length. For that, we use a *zip* function. The general syntax is: *dict(zip(list1,list2))*. It takes two lists as parameters. The first list "*list1*" acts as a key and the second list "*list2*" acts as a value of the dictionary.

Example 6.8

```
keys = [1,2,3,4]
values = ['crow', 'peacock', 'parrot', 'swan']
print("List1 is ", keys)
print("List2 is ", values)
d_bird = dict(zip(keys,values))
print("The dictionary is ")
print(d_bird)
```

Output

```
List1 is [1, 2, 3, 4]
List2 is ['crow', 'peacock', 'parrot', 'swan']
The dictionary is:
{1: 'crow', 2: 'peacock', 3: 'parrot', 4: 'swan'}
```

6.2.1.6 Insertion of a New Key–Value

Adding a new key–value pair to a dictionary is very simple. Assign a value to a new, previously non-existent key to add a new key–value pair to a dictionary. The general syntax is: *dictionary_name[key]=value*.

Example 6.9 Adding item to dictionary

```
Animals = {1:'cat', 2:'cow', 3:'camel'}
Animals[4] = 'cheetah'
print(Animals)
```

In order to add one more animal 'cheetah' to the directory "Animals" assign it to a non-existing key of the dictionary.

Output

```
{1: 'cat', 2: 'cow', 3: 'camel', 4: 'cheetah'}
```

6.2.2 Dictionary Assignment

The assignment of a dictionary can be done by equal to ("=") symbol. This can also be done with the *copy()* function.

Example 6.10

```
my_dict1 = dict(([1,'cat'],[2,'cow'],[3,'camel'],[4,'che
etah']))
my_dict2 = my_dict1
print(my_dict1 == my_dict2)
my_dict3 = my_dict1.copy()
print(my_dict3 == my_dict2)
```

Output

```
True
True
```

6.2.3 Accessing Dictionary

There are various ways we can access the dictionary.

If the key of the dictionary is known, we can find its value. The general syntax is: ***dictionary_Name[key]***. It is the same as accessing an item in a list but the difference is in a list we pass the index whereas in a dictionary we pass the key.

Example 6.11 Access a dictionary item

```
Bird_Sound = {'crow':'caw', 'Doves':'coo', 'Ducks':'quack',
'Owl':'hoot'}
x = Bird_Sound['Doves']
print(x)
```

In order to get the value of *key='Dove'*, use the dictionary name followed by key within square brackets.

Output:

```
coo
```

There are two different methods to access items from a dictionary.

6.2.3.1 keys()

This method returns a list built of all the keys within the dictionary. Having a list of keys enables you to access the whole dictionary easily. The general syntax is: *dictionary_Name.keys()* fetches all the keys from the dictionary.

Example 6.12 Access a dictionary item with keys()

```
Animals = {1:'cat', 2:'cow', 3:'camel', 4:'cheetah'}
for x in Animals.keys():
  print(x, ' -> ', Animals[x])
```

Here x is the iterable which has all the keys of the dictionary Animals. Inside the for-loop we have accessed all the values corresponding to each key.

Output

```
1 -> cat
2 -> cow
3 -> camel
4 -> cheetah
```

6.2.3.2 items()

The method returns a list of tuples, where each tuple is a key–value pair. The general syntax is: *dictionary_Name.items()*.

Example 6.13 Access a dictionary item with items()

```
Animals = {1:'cat', 2:'cow', 3:'camel', 4:'cheetah'}
for x,y in Animals.items():
  print(x, '->', y)
```

Observe that the variables x and y take a different key and value pair from the dictionary. For each key, x, the corresponding value, y is fetched.

Output

```
1 -> cat
2 -> cow
3 -> camel
4 -> cheetah
```

6.2.3.3 get()

This method returns the value which is associated with the key. The syntax is:
dictionary_name.get(key, [default=None]). It takes two parameters. The first parameter is "key", which is to be searched in the directory. If the key exists, it returns the corresponding value. The second parameter is optional, "default=None". It is the value to be returned if the key is not found in the directory.

Example 6.14

```
Employee = {'Name':'Ram','Age':30,'Salary':12500.50}
x = Employee.get('Name')
print(x)
```

Observe that the key passed to *get()* is 'Name' and its corresponding value 'Ram' is returned.

Output

```
Ram
```

Observe that the *dictionary_Name.get(key)* method returns the same as *dictionary_Name[key]*. But there is a difference, the *get()* method returns None for a non-existing key, but the other method returns an error message for a non-existing key.

Example 6.15

```
Employee = {'Name':'Ram','Age':30,'Salary':12500.50}
x = Employee.get('Name')
print(x)
y = Employee['Gender']
print(y)
```

Observe that 'Gender' is a key which is non-existing in the dictionary Employee. The get() method returned None, which is the default value. When Employee['Gender'] is used, Python called Employee and finds a value which is associated with the key Gender. Since the key Gender does not exist, it returns an error.

Output

```
Ram
Traceback (most recent call last):
  File "C:\Users\RakeshN\AppData\Local\Programs\Python\Py
  thon37-32\chap-6.py", line 149, in <module>
  y = Employee['Gender']
KeyError: 'Gender'
```

If the default (optional) parameter is also passed, this parameter is returned if the key is not found in the dictionary.

Example 6.16

```
Employee = {'Name':'Ram', 'Age':30, 'Salary':12500.50}
x = Employee.get('Gender', "No Such key Exist")
print(x)
```

Here for default parameters, *"No Such key Exist"* is passed. When we tried to find the value of a non-existing key *"Gender"*, it returned this parameter.

Output

```
No Such key Exist
```

6.2.3.4   Membership in Dictionary

The key word *in* is a membership function. It checks if a key is in the dictionary or not. It returns a Boolean value. If the given key is in the dictionary, it returns True otherwise it returns False.

Example 6.17

```
my_dict1 = dict({1:'cat', 2:'cow', 3:'camel', 4:'cheetah'})
x = int(input('Enter the key: '))
if x in my_dict1:
  print(x, " exist in the dictionary.")
else:
  print(x, " does not exist in the dictionary.")
```

Output

```
Enter the key: 2
2 exist in the dictionary.
Enter the key: 6
6 does not exist in the dictionary.
```

6.2.4   Dictionary Methods

There are many methods available for dictionaries. The methods like *len()*, *del*, *pop()*, *clear()* have the same syntax and use as in list and String. Few more useful dictionary methods are discussed here.

6.2.4.1   update()

This method updates a dictionary by using another dictionary. The general syntax is: ***dictionary_Name.update(Another_Dictionary_Name)***. It takes a parameter, which is another dictionary. If the keys of both the dictionaries are different, the dictionary that is passed as a parameter is added to another dictionary. If there is a value which has the same key in both these dictionaries, then the key of the dictionary which is getting updated will no longer be there. We cannot have two items in a dictionary with the same key.

Example 6.18 update()

```
Bird_Sound = {'Crow':'caw','Dove':'coo'}
Animals = {1:'cat', 2:'cow', 3:'camel', 4:'cheetah'}
```

```
print('\nDictionary before update() ')
print('Bird_Sound =', Bird_Sound)
print('Animals = ', Animals)
Animals.update(Bird_Sound)
print('\nDictionary after update() ')
print('Bird_Sound = ', Bird_Sound)
print('Animals = ', Animals)
```

Observe that after applying *update()*, the *Bird_Sound* dictionary remains the same, whereas the *Animals* dictionary is updated. In the updated dictionary the original key–value pair of Animals appeared first and then the key–value pair of Bird_Sound appeared. This basically means that it takes every object in the *Bird_ Sound* and adds it to the dictionary Animals. So the dictionary Animals is changed.

Output

```
Dictionary before update()
Bird_Sound = {'Crow': 'caw', 'Dove': 'coo'}
Animals = {1: 'cat', 2: 'cow', 3: 'camel', 4: 'cheetah'}
Dictionary after update()
Bird_Sound = {'Crow': 'caw', 'Dove': 'coo'}
Animals = {1: 'cat', 2: 'cow', 3: 'camel', 4: 'cheetah',
'Crow': 'caw', 'Dove': 'coo'}
```

Example 6.19 update()

```
Bird = {1:'Crow',2:'Dove', 3:'Owl'}
Animals = {1:'Cat', 2:'cow', 3:'camel', 4:'Cheetah'}
print('\nDictionary before update()')
print('Bird =', Bird)
print('Animals = ',Animals)
Animals.update(Bird)
print('\nDictionary after update()')
print('Bird =', Bird)
print('Animals = ', Animals)
```

Observe that here the *Animals* directory is updated by the *Bird* directory. Both the directories have some common keys (i.e., 1, 2, 3). In the Animals directory, values with the same keys will be getting updated, So 'cat', 'cow' and 'camel' will be overwritten by 'Crow', 'Dove' and 'Owl'. We cannot have two items in a diction-ary with the same key.

Output

```
Dictionary before update()
Bird = {1: 'Crow', 2: 'Dove', 3: 'Owl'}
Animals = {1: 'Cat', 2: 'cow', 3: 'camel', 4: 'Cheetah'}
Dictionary after update()
Bird = {1: 'Crow', 2: 'Dove', 3: 'Owl'}
Animals = {1: 'Crow', 2: 'Dove', 3: 'Owl', 4: 'Cheetah'}
```

6.2.4.2   sorted()

It sorts the dictionary in terms of the keys. One more syntax is: *sorted(dictionary_name).*

Example 6.20

```
Employee = {'Name':'Ram', 'Age':30, 'Salary':12500.50}
print("Orignal Dictionary")
for x in Employee.keys():
  print(x,' -> ', Employee[x])
print("\nSorted Dictionary")
for x in sorted(Employee.keys()):
  print(x,' -> ', Employee[x])
```

Observe the 6th line, we have used *sorted(Employee.keys())* for sorting. The given dictionary is sorted according to the key.

Output

```
Orignal Dictionary
Name -> Ram
Age -> 30
Salary -> 12500.5
Sorted Dictionary
Age -> 30
Name -> Ram
Salary -> 12500.5
```

6.2.4.3   str()

This function returns a String of a given dictionary. The syntax is: *str(dictionary_Name).* The parameter passed is the name of the dictionary.

Example 6.21

```
Employee = {'Name':'Ram',"Age":30,'Salary':12500.50}
str1 = str(Employee)
print(str1)
```

Output

```
{'Name': 'Ram', 'Age': 30, 'Salary': 12500.5}
```

6.2.4.4   setdefault()

This function is used to set a value of a key. The general syntax is: *setdefault(key,[default_value=None]).* It takes two parameters, the key and the default value, in which the default value is optional. If the key is present, the default value is set for the key. If the key is absent, the key along with the default value is added to the dictionary. If the default value is missing, it takes *None* as the default value.

Example 6.22

```
Emp1 = {'Name':'Ram','Age':30, 'Salary':12500.50}
Emp1.setdefault('Gender')
print("Key Missing", Emp1)
Emp1= {'Name':'Ram','Age':30,'Salary':12500.50}
Emp1.setdefault('Gender','Male')
print("Key Present", Emp1)
```

Output

```
Key Missing {'Name': 'Ram', 'Age': 30, 'Salary': 12500.5,
'Gender': None}
Key Present {'Name': 'Ram', 'Age': 30, 'Salary': 12500.5,
'Gender': 'Male'}
```

WORKED OUT EXAMPLES

Example 6.23 Write a program in Python to create a new dictionary for a given dictionary by reversing the key–value pair.

```
Employee = {'Name':'Ram','Age':30,'Salary':12500.00}
new_Emp={}
print("Original Dictionary")
for x in Employee.keys():
  print(x, "->", Employee[x])
  new_Emp[Employee[x]]=x

print("\nNew Dictionary")
for x in new_Emp.keys():
print(x,"->",new_Emp[x])
```

Output

```
Original Dictionary
Name -> Ram
Age -> 30
Salary -> 12500.0
New Dictionary
Ram -> Name
30 -> Age
12500.0 -> Salary
```

Example 6.24 Write a program that reads data from a keyboard to create a dictionary.

```
number = int(input("How many key-value pair to enter? "))
my_dict = {}
for x in range(number):
  k1= eval(input("Enter the Key: "))
```

```
v1= eval(input("Enter the Value: "))
my_dict[k1] = v1
print("The dictionary is ", my_dict)
```

Output

```
How many key-value pair to enter? 4
Enter the Key: 1
Enter the Value: 'Rose'
Enter the Key: 2
Enter the Value: 'Lily'
Enter the Key: 3
Enter the Value: 'Lotus'
Enter the Key: 4
Enter the Value: 'Jasmine'
The dictionary is {1: 'Rose', 2: 'Lily', 3: 'Lotus', 4:
'Jasmine'}
```

MULTIPLE CHOICE QUESTIONS

1. Find the output of the following code
   my_dict = {1:0,2:3,3:1,0:2}
   x = 0
   for y in range(len(my_dict)):
       x = my_dict[x]
   print(x)

   a. 0
   b. 1
   c. 2
   d. 3

2. Find the output of the following code
   d1 = {}
   d1['2'] = [2,1]
   d1['1'] = [4,3]
   for x in d1.keys():
       print(d1[x][1],end ="")

   a. 12
   b. 13
   c. 14
   d. 15

3. Find the output of the following code
   d1 = {'one':1,'three':3,'two':2}
   for k in sorted(d1.values()):
       print(k,end="")
```
```

   a. 321
   b. 312
   c. 123
   d. 132

4. Find the code for which the output is abc

```
d1 = { }
l1 = ['a', 'b', 'c','d']
for i in range (len(l1) –1):
    d1[l1[i]] = (l1[i],)
for i in sorted(d1.keus()):
   k = d1[i]
   # Your Code
```

   a. print(k[0])
   b. print(k["0"])
   c. print(k['0'])
   d. print(k)

5. Find the output of the following code

```
d1 = {'one':'two','three':'one','two':'three'}
v = d1['one']
for k in range(len(d1)):
    v = d1[v]
print(v)
```

   a. two
   b. three
   c. one
   d. None

6. Find the output of the following code

```
d1 = {1:'one',2:'two'}
d2 = {1:'one',2:'two'}
print(d1 == d2)
print(d1 is d2)
```

   a. False.True
   b. True,True
   c. False,False
   d. True,False

7. Find the output of the following code

```
d1 = {1:'one',2:'two'}
d2 = {2:'one',3:'two'}
print(d1 > d2)
```
   a. True
   b. False

    c. Error

    d. None of the above

8. Which command to use to delete the entry 2 from the code.

   d1 = {1:'one',2:'two'}

    a. d1.delete(2:'two')

    b. d1.delete(2)

    c. del d1(2)

    d. del d1('two')

9. Find the output of the following code

   d1 = {1:'one',2:'two'}

   print(d1[3])

    a. ValueError

    b. AttributeError

    c. KeyError

    d. SyntaxError

10. Find which of the following statement is false

    a. Key can be access from value

    b. Value can be accessed from key

    c. Dictionary is a unordered

    d. Dictionary is mutable

11. Find which of the following statement is incorrect

    a. {1:2, 2:3, 3:4}

    b. dict([[1,2],[2,3],[3,4]])

    c. { }

    d. {1=2,2=3,3=4}

12. Find the output of the code

   d1 = {1:'one',2:'two'}

   for x,y in d1.items():

      print(x,y,end="")

    a. 1 2 one two

    b. 2 two1 one

    c. 1 one2 two

    d. None of the above

13. Find the output of the code

   d1 = {1:'one',2:'two'}

   print(d1.get(1,3))

    a. one

    b. two

    c. 1

    d. 2

14. Find the output of the code
    d1 = {1:'one',2:'two'}
    print(d1.get(3,4))
    a. one
    b. two
    c. 3
    d. 4

15. Find the output of the code
    d1 = {1:'one',2:'two'}
    d1.setdefault(3,'three')
    print(d1)

    a. {1: 'one', 2: 'two', 3: 'three'}
    b. {1: 'three', 2: 'three', 3: 'three'}
    c. {3: 'one', 3: 'two', 3: 'three'}
    d. None of the above

16. Find the output of the code
    d1 = {1:'one',2:'two'}
    d1 = {2:'one',3:'two'}
    d1.update(d2)
    print(d1)

    a. {3: 'one', 4: 'two',1: 'one', 2: 'two' }
    b. {1: 'one', 2: 'two', 3: 'one', 4: 'two'}
    c. {1: 'one', 2: 'one', 3: 'two'}
    d. None of the above

17. Find the output of the code
    d1 = {1:'one',2:'two'}
    d1.clear()
    print(d1)

    a. { }
    b. {1: 'None', 2: 'None'}
    c. {None: ' None ', None: ' None '}
    d. None of the above

18. Find the output of the code
    d1 = {1:'one',2:'two'}
    d1.pop('two')
    a. ValueError
    b. AttributeError
    c. KeyError
    d. None of the above

19. Find the output of the code
```
d1 = {1:'one',2:'two'}
x = d1.pop(3,4)
print(x)
```
    a. ValueError
    b. 4
    c. 3
    d. None of the above

20. Find the output of the code
```
d1 = {}
d1[1] = 1
d1['1'] = 2
d1[1.0] = 4
count = 0
for i in d1:
    count += d1[i]
print(count)
```
    a. 3
    b. 4
    c. 4
    d. 6

DESCRIPTIVE QUESTIONS

1. What are the different ways of creating an array?
2. What is the difference between dict.get() and dict[key]?
3. Explain the setdefault() method in dictionary.
4. What is the difference between pop, del and clear methods/functions in dictionary?
5. What is hashing? Explain hashable object and their hash values in dictionary.

PROGRAMMING QUESTIONS

1. Write a program that prints all the unique values of a dictionary.
2. Write a program to check if a dictionary is empty or not.
3. Write a program to remove duplicate values from Dictionary.
4. Write a program that finds the sum of all keys and all values. Assume that the keys and values are numeric.
5. Write a script to print a dictionary where the keys are numbers between 1 and 15 (both included) and the values are squares of keys.
6. Write a program to sort a list alphabetically in a dictionary.

ANSWER TO MULTIPLE CHOICE QUESTIONS

1.	A	2.	B	3.	C	4.	A	5.	A
6.	D	7.	C	8.	C	9.	C	10.	A
11.	D	12.	C	13.	A	14.	D	15.	A
16.	C	17.	A	18.	C	19.	B	20.	D

7 Set

LEARNING OBJECTIVES

After studying this chapter, the reader will be able to:

- Create/initialize a set
- Access elements in a set
- Perform different set operations
- Use some common functions as well as methods in a set
- Understand frozen set

7.1 INTRODUCTION

Set is yet another data structure which is non-sequential. The elements are non-repeating. The concept of set theory in Mathematics is the same in Python. Set has a highly optimized method for checking whether an element belongs to it or not.

7.2 SET

A Set is an unordered and unindexed collection of unique elements. A set in Python is defined as a collection of elements in curly brackets.

Set_Variable_Name = { item1, item2, item3,…,itemn}

Where items can be a single item or a list or one more tuple.
In a set notice that:

- A set *starts* and *ends* with a pair of *curly brackets { }.*
- The items that are filled within the curly brackets are separated by commas.
- The items inside a list may have *different types.*
- Set can be a collection of strings, numeric values and tuples.
- A set can also have a keyword set followed by a collection of items.
- The items are non-repeating in a set.

Example 7.1

```
Set_A= {'Crow','Peacock','Owl',"Parrot"}
print(Set_A)
```

Observe that the items inside the curly braces are strings.

DOI: 10.1201/9781003219125-8

Output:

```
{'Peacock', 'Parrot', 'Owl', 'Crow'}
```

A set may contain a collection of strings, numbers and tuples or a combination of all of these.

Example 7.2

```
Set_A= {'Crow', ('Owl', 'Parrot'), 5.2}
print(Set_A)
```

Observe that there is a tuple as an item of the set.

Output

```
{'Crow', ('Owl', 'Parrot'), 5.2}
```

Example 7.3

```
Set_A= set(['Crow','Peacock','Owl',"Parrot",'Crow'])
print(Set_A)
print(type(Set_A))
```

Observe that a set is created by the keyword *set*. Set is type cased on a list. Also observe that there are two entries of 'Crow' in the definition of the set, but in the output, there is only one 'Crow'. This implies the items in the set are non-repeated or unique.

Output

```
{'Crow', 'Owl', 'Peacock', 'Parrot'}
<class 'set'>
```

7.2.1   CREATING A SET

7.2.1.1   Creating an Empty Set

Recall that Directories are also created by curly brackets. Empty curly brackets { } will create an empty Directory. The syntax for creating a empty set is: *Set_Name = set()*.

Example 7.4

```
A = set()
print(A)
print(type(A))
```

Observe that in order to create an empty set, the keyword *set* is used followed by a pair of braces ( ).

Output

```
set()
<class 'set'>
```

7.2.1.2   Creating a Set from a List, Tuples and String

We can create a set from a list or a tuple or a String by type casting set( ).

Example 7.5

```
lst= [1, 2, 3, 4, 5]
set_A=set(lst)
print("The elements are ", set_A, " the type is ",
type(set_A))
tup= (1,2,3,4,5)
set_B= set(tup)
print("The elements are ", set_B, " the type is ",
type(set_B))
st= "abcdef"
set_C=set(st)
print("The elements are ", set_C, " the type is ",
type(set_C))
```

Here a list *lst,* a tuple *tup* and a String *st* are converted to a set by type casting *set.* Each element in these data structures becomes an element in the set.

Output

```
The elements are {1, 2, 3, 4, 5} the type is <class 'set'>
The elements are {1, 2, 3, 4, 5} the type is <class 'set'>
The elements are {'a', 'e', 'f', 'd', 'c', 'b'} the type is
<class 'set'>
```

7.2.1.3   Creating a Set from a Dictionary

A set can be created from a dictionary also by type casting but the keys only (not the values) become elements of the set.

Example 7.6

```
D1= {1:"a",2:55,4:"def"}
set_D = set(D1)
print("The elements are ", set_D, " the type is ",
type(set_D))
```

In a dictionary, the elements are represented in the form of a key–value pair. While converting it to a set we need to perform type casting. Only the key of the dictionary forms a set.

Output

```
The elements are {1, 2, 4} the type is <class 'set'>
```

7.2.2   ACCESSING SET ELEMENTS

As set is an unordered collection of elements, we cannot access the set elements with an index. However, we can traverse the elements of a set using a looping structure.

Example 7.7

```
Set_A= {1, 2, 2, 3, 4, 5, 6}
print("The set is ", Set_A)
print("The elements in the set are ")
for i in Set_A:
  print(i)
```

Output

```
The set is {1, 2, 3, 4, 5, 6}
The elements in the set are
1
2
3
4
5
6
```

7.2.2.1   Membership in Set

The keyword *in* is a membership function. It checks if an element is in the set or not. It returns a Boolean value. If the given element is in the set, it returns True otherwise it returns False.

Example 7.8

```
Set_A= set([1, 2, 3, 4, 5, 6, 8, 10])
i = 5
if i in Set_A:
  print(i," is an element of the set ", Set_A)
else:
  print(i," is an NOT element of the set ", Set_A)
```

Observe that 5 is an element in Set_A. So, it returned TRUE.

Output

```
5 is an element of the set {1, 2, 3, 4, 5, 6, 8, 10}
```

7.2.3   SET OPERATIONS

There are several operations available for sets. The operations available in mathematical concepts of set theory are also available.

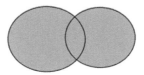

FIGURE 7.1 Union of sets

7.2.3.1 union()

This method returns a set of all those elements which are in either of the sets as shown in figure 7.1. The general syntax is: *set1.union(set2) or set1 | set2*. It takes two parameters as two sets.

Example 7.9

```
Set_Birds= set(['Crow','Peacock','Owl','Parrot'])
Set_Animals= set(['Dog','Cat','Pig'])
print("Set before union of sets \n", Set_Birds, Set_Animals)
Set_Domestic= Set_Birds.union(Set_Animals)
print("Set after union of sets \n", Set_Domestic)
```

Example 7.10

```
Set_Birds= set(['Crow','Peacock','Owl','Parrot'])
Set_Animals= set(['Dog','Cat','Pig'])
print("Set before union of sets \n", Set_Birds, Set_Animals)
Set_Domestic= Set_Birds | Set_Animals
print("Set after union of sets \n", Set_Domestic)
```

Observe that in this example 7.9 the union() method is used and in example 7.10, the "|" symbol is used for the set union. The output remains the same.

Output

```
Set before union of sets
{'Parrot', 'Peacock', 'Crow', 'Owl'} {'Cat', 'Pig', 'Dog'}
Set after union of sets
{'Peacock', 'Owl', 'Pig', 'Parrot', 'Cat', 'Crow', 'Dog'}
```

7.2.3.2 intersection()

This method returns all those elements which are in both sets as shown in figure 7.2. The general syntax is: *Set1.intersection(Set2) or Set1 & Set2*. It takes two parameters as two sets.

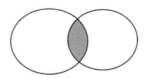

FIGURE 7.2 Intersection of sets

Example 7.11

```
Set_A= set([1, 2, 3, 4, 5, 6])
Set_B= set([2, 4, 6, 8])
print("Set before intersection of sets \n", Set_A,Set_B)
Set_C= Set_A.intersection(Set_B)
print("Set after intersection of sets \n", Set_C)
```

Example 7.12

```
Set_A= set([1, 2, 3, 4, 5, 6])
Set_B= set([2, 4, 6, 8])
print("Set before intersection of sets \n", Set_A,Set_B)
Set_C= Set_A & Set_B
print("Set after intersection of sets \n", Set_C)
```

Observe that in this example 7.11 the intersection() method is used and in example 7.12, the "&" symbol is used for the set intersection. The output remains the same.

Output

```
Set before intersection of sets
{1, 2, 3, 4, 5, 6} {8, 2, 4, 6}
Set after intersection of sets
{2, 4, 6}
```

7.2.3.3 difference()

This method returns all those elements which are in the first set but not in the second set as shown in figure 7.3. The general syntax is: ***set1.differencce(Set2) or Set1 – Set2***.

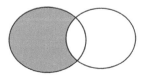

FIGURE 7.3   Difference of sets

Example 7.13

```
Set_A= set([1, 2, 3, 4, 5, 6])
Set_B= set([2, 4, 6, 8])
print("Set before difference of sets \n", Set_A,Set_B)
Set_C= Set_A.difference(Set_B)
print("Set after difference of sets \n", Set_C)
```

Example 7.14

```
Set_A= set([1, 2, 3, 4, 5, 6])
Set_B= set([2, 4, 6, 8])
print("Set before difference of sets \n", Set_A, Set_B)
Set_C= Set_A - Set_B
print("Set after difference of sets \n", Set_C)
```

Observe that in this example 7.13 the difference() method is used and in example 7.14, "–" symbol is used for the set difference. The output remains the same.

Output

```
Set before difference of sets
{1, 2, 3, 4, 5, 6} {8, 2, 4, 6}
Set after difference of sets
{1, 3, 5}
```

7.2.3.4   Exclusive-OR

This operation returns all those elements which are either in one set or the other set but not in both as shown in figure 7.4. The general syntax is: *set1 ^ set2*.

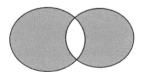

FIGURE 7.4   Exclusive-OR of sets

Example 7.15

```
Set_A= set([1, 2, 3, 4, 5, 6])
Set_B= set([2, 4, 6, 8])
print("Set before Exclusive-OR of sets \n", Set_A, Set_B)
Set_C= Set_A ^ Set_B
print("Set after Exclusive-OR of sets \n", Set_C)
```

Observe the output, the elements {1, 3, 5} form Set_A and the element {8} from Set_B are printed. All the elements common to Set_A and Set_B are not a part of the output.

Output

```
Set before Exclusive-OR of sets
{1, 2, 3, 4, 5, 6} {8, 2, 4, 6}
Set after Exclusive-OR of sets
{1, 3, 5, 8}
```

7.2.3.5   Symmetric Difference

This method returns all those elements which are in set1 but not in set2 union all those elements which are in set2 but not in set1 in other words *(set1 Union set2) – (set2 Union Set1)* as shown in figure 7.4. The general syntax is: *set1.symmetric_difference(Set2)*.

Example 7.16

```
Set_A= {1, 2, 3, 4, 5, 6}
print("The set A is ", Set_A)
Set_B= {2, 4, 6, 8, 10}
print("The set A is ", Set_B)
Set_C= Set_A.symmetric_difference(Set_B)
print("The Symmetric Difference set of set A and B is ",
Set_C)
```

Observe the output it returned the union of {1, 3, 5} from set1 which is not in set2 and {8, 10} from set2 which are not in set1.

Output

```
The set A is {1, 2, 3, 4, 5, 6}
The set A is {2, 4, 6, 8, 10}
The symmetric difference set of set A and B is {1, 3, 5, 8, 10}
```

7.2.3.6   Equivalent Set and Not Equivalent

This operation returns a Boolean variable, it checks if two sets are the same or not. The general syntax is: *Set1 == Set2*. If the two sets are equivalent, it returns False.

The general syntax to check not equivalence is: *Set1 != Set2*.

Example 7.17

```
Set_A= set([1, 2, 3, 4, 5, 6, 8])
Set_B= set([2, 4, 6, 8])
Set_C= {5, 8, 2, 6, 3, 4, 1}
if Set_A == Set_C:
  print("Set_A and Set_C are equivalent ")
else:
print("Set_A and Set_C are not equivalent ")
if Set_B != Set_C:
  print("Set_B and Set_C are not equivalent")
else:
  print("Set_B and Set_C are equivalent")
```

Observe *Set_A* and *Set_C*, both the sets are defined in two different ways with different ordering. It returned *True* for *Set_A == Set_C* and returned *False* for *Set_B != Set_C*.

Output

```
Set_A and Set_C are equivalent
Set_B and Set_C are not equivalent
```

7.2.3.7 Subset and Superset

The concepts of subset and superset in Python are the same as in Mathematics. When all the elements of Set_A are a part of Set_B or Set_A is equal to Set_B then, Set_A is a subset of Set_B or Set_B is a superset of Set_A, as shown in figure 7.6.

The general syntax for subset is: ***Set1 < = Set2***.

The general syntax for superset is: ***Set1 > = Set2***.

This returns a Boolean variable. If Set_A is a subset or equal to Set_B, then it returns True otherwise it returns False. If Set_B is a superset or equal to Set_A, then it returns True otherwise it returns False (figure 7.5).

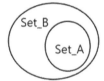

FIGURE 7.5 Subset and superset

Example 7.18

```
Set_A= set([1, 2, 3, 4, 5, 6, 8, 10])
Set_B= set([2, 4, 6, 8])
Set_C= {1, 2, 3, 4, 5, 6, 8, 10}
if Set_B <= Set_A:
  print("Set_B is subset of Set_A.")
else:
  print("Set_B is not subset of Set_A.")
if Set_A >= Set_B:
  print("Set_A is a superset of Set_B.")
else:
  print("Set_A is not superset of Set_B.")
if Set_A <= Set_C:
  print("Set_A is subset of Set_C.")
else:
  print("Set_A is Not subset of Set_B")
```

All the elements of Set_B are in Set_A. So Set_B is a subset of Set_A. Similarly, Set_A is a superset of Set_B. Observe that Set_A and Set_C are the same. So Set_A is either subset or superset of Set_B.

Output

```
Set_B is subset of Set_A.
Set_A is a superset of Set_B.
Set_A is subset of Set_C.
```

7.2.3.8   Proper Subset and Proper Superset

The concepts of proper subset and proper superset in Python are the same as in Mathematics. When all the elements of Set_A are a part of Set_B, Set_A is a proper subset of Set_B or Set_B is a proper superset of Set_A, as shown in figure 7.7.

The general syntax for a proper subset is: ***Set1 < Set2***.

The general syntax for proper superset is: ***Set1 > Set2***.

This returns a Boolean variable. If Set_A is a proper subset of Set_B, then it returns True otherwise it returns False. If Set_B is a proper superset of Set_A, then it returns True otherwise it returns False.

Example 7.19

```
Set_A= set([1, 2, 3, 4, 5, 6, 8, 10])
Set_B= set([2, 4, 6, 8])
Set_C= {1, 2, 3, 4, 5, 6, 8, 10}
if Set_B < Set_A:
  print("Set_B is proper subset of Set_A.")
else:
  print("Set_B is not proper subset of Set_A.")
if Set_A > Set_B:
  print("Set_A is a proper superset of Set_B.")
else:
  print("Set_A is not proper superset of Set_B.")
if Set_A > Set_C:
  print("Set_A is not proper superset of Set_C.")
else:
  print("Set_A is proper superset of Set_C")
```

All the elements of Set_B are in Set_A. So Set_B is a proper subset of Set_A. Similarly, Set_A is a proper superset of Set_B. Observe that Set_A and Set_C are the same. So neither Set_A is a proper subset nor a proper superset of Set_C. Set_A is either subset or proper superset of Set_C.

Output

```
Set_B is proper subset of Set_A.
Set_A is a proper superset of Set_B.
Set_A is proper superset of Set_C
```

7.2.4   SET FUNCTION AND METHODS

Like any other data structure, set has many methods such as add element, delete element and access element. Some functions and usage such as *len()*, *update()*, *clear()*, *remove()*, *pop() and del()* are the same as in list and String. We are going to discuss some more methods and operations here.

7.2.4.1   add()

This method adds an element to the existing set. The general syntax is: **Set_name. add(element)**. A parameter as an element to be inserted is passed. The value that is passed to this function is added to the existing set.

Example 7.20

```
Set_A= set(['Crow', 'Peacock', 'Owl', 'Parrot'])
print("Original set\n", Set_A)
Set_A.add('Pigeon')
print("Set after addition of elemet\n", Set_A)
```

Observe that a new element *'Pigeon'* is added to the set by passing this value as parameter to *Set_A.add()*.

Output

```
Original set
{'Peacock', 'Maina', 'Parrot'}
Set after update of element
{'Crow', 'Peacock', 'Maina', 'Sparrow', 'Pigeon', 'Parrot'}
```

7.2.4.2   discard()

This method deletes an element from the set. But it does not raise an error if the element is not in the set. The general syntax is: **Set_name.discard(element)**. It takes a parameter as an element that is to be discarded.

Example 7.21

```
Set_A= {'Peacock', 'Maina', 'Parrot'}
print("The original set is ", Set_A)
rem= input("Enter the element to be discarded: ")
Set_A.discard(rem)
print("After discard of", rem,"the set is", Set_A)
```

Output

```
The original set is {'Peacock', 'Maina', 'Parrot'}
Enter the element to be discarded: Maina
After discard of Maina the set is {'Peacock', 'Parrot'}
```

Output

```
The original set is {'Peacock', 'Parrot', 'Maina'}
Enter the element to be discarded: Crow
After discard of Crow the set is {'Peacock', 'Parrot',
'Maina'}
```

The function remove() removes the element if it is present in the set. If the element is not present, it throws an error. The function discard() removes the element if it is present in the set. If the element is not present, no action is performed (Error is not thrown).

7.2.4.3   isdisjoint()

This method checks if any common elements are between two sets. In other words, if the intersection of two sets has some elements, they are not disjoint. The general syntax is: *Set1.isdisjoint(Set2)*. It returns True if there are no common elements, otherwise returns False.

Example 7.22

```
Set_A= {'Peacock', 'Maina', 'Parrot'}
Set_B= {'Crow', 'Owl',"Sparrow"}
Set_C= {'Owl', "Crow", 'Parrot'}
print("The Set_A is", Set_A)
print("The Set_B is", Set_B)
print("The Set_C is", Set_C)
print("The two sets Set_A and Set_B disjoint?", Set_A.
isdisjoint(Set_B))
print("The two sets Set_A and Set_B disjoint?", Set_A.
isdisjoint(Set_C))
```

Output

```
The Set_A is {'Peacock', 'Maina', 'Parrot'}
The Set_B is {'Sparrow', 'Owl', 'Crow'}
The Set_C is {'Parrot', 'Owl', 'Crow'}
The two sets Set_A and Set_B disjoint? True
The two sets Set_A and Set_B disjoint? False
```

7.2.4.4   issubset()

This method checks if any one set is a subset of other sets. The general syntax is: *Set1.issubset(Set2)*. It returns True if Set1 is a subset of Set2, otherwise returns False.

Example 7.23

```
Set_A= {'Peacock', 'Maina',"parrot", 'Crow', 'Owl',
"Sparrow"}
Set_B= {'Crow', 'Owl', 'Sparrow'}
print("The Set_A is", Set_A)
print("The Set_B is", Set_B)
print("The Set_A is subset of Set_B?", Set_A.
issubset(Set_B))
print("The Set_B is subset of Set_A?", Set_B.
issubset(Set_A))
```

Output

```
The Set_A is {'Owl', 'Crow', 'Peacock', 'Sparrow', 'Maina',
'parrot'}
The Set_B is {'Sparrow', 'Owl', 'Crow'}
The Set_A is subset of Set_B? False
The Set_B is subset of Set_A? True
```

7.2.4.5   issuperset()

This method checks if any one set is a superset of other sets. The general syntax is: *Set1.issuperbset(Set2)*. It returns True if Set1 is a superset of Set2, otherwise returns False.

Example 7.24

```
Set_A= {'Peacock', 'Maina', 'Parrot', 'Crow', 'Owl',
'Sparrow'}
Set_B= {'Crow', 'Owl', 'Sparrow'}
print("The Set_A is", Set_A)
print("The Set_B is", Set_B)
print("The Set_A is superset of Set_B?", Set_A.
issuperset(Set_B))
print("The Set_B is superset of Set_B?", Set_B.
issuperset(Set_A))
```

Output

```
The Set_A is {'Parrot', 'Maina', 'Owl', 'Crow', 'Peacock',
'Sparrow'}
The Set_B is {'Owl', 'Crow', 'Sparrow'}
The Set_A is superset of Set_B? True
The Set_B is superset of Set_B? False
```

7.2.4.6   Frozen Set

When the elements of the set are immutable, such sets are called "*Frozen set*". The general syntax is: *Set_Variable_Name = frozenset([item1,item2....,itemn])*.

Example 7.24

```
Set_A= frozenset(['Crow',5.2,('Owl', 'Parrot')])
print(Set_A)
print(type(Set_A))
```

In order to create a frozen set, the keyword frozen is used on list.

Output

```
frozenset({('Owl', 'Parrot'), 'Crow', 5.2})
<class 'frozenset'>
```

Example 7.25

```
Set_A= frozenset([1, 2, 3, 4, 5, 6, 8, 10])
Set_A.add(11)
print("Addition of an element to frozen set", Set_A)
```

As frozen sets are immutable, when manipulation on data is tried, it returns an Attribute Error.

Output

```
Traceback (most recent call last):
  File "C:\Users\RakeshN\AppData\Local\Programs\Python\Py
  thon37-32\chap-7.py", line 400, in <module>
  Set_A.add(11)
AttributeError: 'frozenset' object has no attribute 'add'
```

There are two major demerits of Python sets:

1. The set doesn't maintain elements in any particular order.
2. Only instances of immutable types can be added to a Python set.

WORKED OUT EXAMPLES

Example 7.26 if A={1,3,5,7,9} and B={2,4,6,8} show that A ∪ (A ∩ B) = A.

```
A= {1, 3, 5, 9}
B= {2, 4, 6, 8}
C= A & B
D= A | C
print(D)
```

Output

```
{1, 3, 5, 9}
```

Example 7.27 if A= {1, 3, 5, 7, 9, 11, 13, 15, 17, 21, 23, 25}. Find a set with elements divisible by 3.

```
A= {1, 3, 5, 7, 9,11, 13, 15, 17, 21, 23, 25}
B= set()
for i in A:
  if i%3==0:
    B.add(i)
print("The original set", A)
print("Set with elements divisible by 3 ->", B)
```

Output

```
The original set {1, 3, 5, 7, 9, 11, 13, 15, 17, 21, 23, 25}
Set with elements divisible by 3 -> {9, 3, 21, 15}
```

Example 7.28 if A={1, 3, 5, 7, 9, 11, 13, 15, 17, 21, 23, 25}. Find a set with elements divisible by 3 or 5.

```
A= {1, 3, 5, 7, 9, 11, 13, 15, 17, 21, 23, 25}
B= set()
for i in A:
  if i%3==0 or i%5==0:
    B.add(i)
print("The original set", A)
print("Set with elements divisible by 3 or 5", B)
```

Output

```
The original set {1, 3, 5, 7, 9, 11, 13, 15, 17, 21, 23, 25}
Set with elements divisible by 3 or 5 {3, 5, 9, 15, 21, 25}
```

MULTIPLE CHOICE QUESTIONS

1. Find the output of the following code
   set1 = {1,2,3,4,3,2,1}
   print(len(set1))
   a. 1
   b. 2
   c. 3
   d. 4

2. Which command will create an empty String?
   a. { }
   b. set( )
   c. [ ]
   d. ( )

3. Find the output of the following code
   set1 = {1, 2, 3, 4, 3, 2, 1}
   set2 = {2, 3, 4}
   print(set1 < set2)

   a. TRUE
   b. FALSE
   c. True
   d. False

4. Find the output of the following code
   set1 = {1, 2, 3, 4}
   set1.add(2)
   print(set1)

   a. {1, 2, 3, 4}
   b. {1, 2, 2, 3, 4}
   c. All of the above
   d. None of the above

5. Find the output of the following code
```
set1 = {1, 2, 3, 4}
set2 = {2, 3, 4, 5}
print(set1 ^ set2)
```

   a. {1, 2, 3, 4, 5}
   b. {1}
   c. {1, 5}
   d. {5}

6. Find the output of the following code
```
set1 = {1, 2, 3, 4}
print(set1 * 3)
```

   a. {1, 2, 3, 4, 1, 2, 3, 4, 1, 2, 3, 4}
   b. {1, 2, 3, 4}
   c. TypeError
   d. ValueError

7. Find the output of the following code
```
set1 = {1, 2, 3, 4}
set2 = {4, 5, 6}
print(set1 | set2)
```

   a. {1, 2, 3, 4, 4, 5, 6}
   b. {1, 2, 3, 4}
   c. {4, 5, 6}
   d. Error

8. Find the output of the following code
```
set1 = {1, 2, 3, 4, {5,6}}
print(set1[3][1])
```

   a. unsupported operand
   b. unhashable type
   c. Name is not defined
   d. None of the above

9. Find which is not correct about frozenset
   a. Immutable
   b. Unordered
   c. Mutable
   d. None of the above

10. Find the output of the following code
```
set1 = frozenset([1, 2, 3, 4])
set1.add(5)
print(set1)
```

    a.   AttributeError
    b.   ValueError
    c.   TypeError
    d.   {1, 2, 3, 4, 5}

11. Find the output of the following code
```
set1 = {1, 2, 3, 4}
set2 = {3, 4, 5, 6, 7}
set1.update(set2)
print(set2)
```

    a.   {1, 2, 3, 4}
    b.   {1, 2, 3, 4, 5, 6, 7}
    c.   {3, 4, 5, 6, 7}
    d.   {4, 5, 6, 7}

12. Find the output of the following code
```
set1 = {1, 2, 3, 4}
set2 = set1
set.remove(2)
print(set2)
```

    a.   {1, 3, 4}
    b.   {1, 2, 3, 4}
    c.   = operation not allowed in set
    d.   {2}

13. Find the output of the following code
```
set1 = {1, 2, 3, 4}
set2 = set1.add(5)
print(set1)
```

    a.   {1, 2, 3, 4, 5}
    b.   {1, 2, 3, 4}
    c.   = operation not allowed in set
    d.   None

14. Find the output of the following code
```
set1 = {1, 2, 3, 4}
print(sum(set1, 5))
```

    a.   5
    b.   10
    c.   15
    d.   50

15. Find the output of the following code
```
set1 = {1, 2}
set2 = {x*x for x in set1 | {5, 6}}
print(set2)
```

    a.   {1, 4}
    b.   {1, 4, 25}
    c.   {1, 4, 25, 36}
    d.   {25, 36}

16. Find the output of the following code

    ```
    set1 = {1, 2}
    print(set1.issubset(set1))
    ```

    a.   True
    b.   False
    c.   true
    d.   false

17. Which method returns a value
    a.   del
    b.   pop
    c.   remove
    d.   clear

18. Which method returns a value

    ```
    set1 = {1, 2, 3, 4}
    set2 = {3, 4, 5, 6}
    print(set1.symmetric_difference(set2))
    ```

    a.   {1, 2, 3, 4}
    b.   {3, 4}
    c.   {5, 6}
    d.   {1, 2, 5, 6}

19. Which one is correct for a proper superset?
    a.   set1>set2
    b.   set1>=set2
    c.   set1<set2
    d.   set1<=set2

DESCRIPTIVE QUESTIONS

1. What are the different ways of creating a set? Explain with examples.
2. Describe different set operations.
3. What are the differences between remove(), del, discard() and clear()?
4. What is a frozen set? Describe an example.
5. What do you mean by shallow copy? Describe with an example.

PROGRAMMING QUESTIONS

1. Write a program that performs symmetric difference operations without using the built-in function.

2. Write a program that performs union operations without using the built-in function.
3. Write a program that performs intersection operations without using the built-in function.
4. Write a program that performs set difference operations without using the built-in function.
5. Write a program that checks if set1 is a subset of set2 without using the built-in function.
6. Write a program to find the length of a set without using the built-in function.
7. Write a program to clear the set without using the built-in function.
8. Write a program to copy a set without using the built-in function.
9. Write a program to check if two sets are equivalent without using the built-in function.
10. Write a program to check if two sets are disjoint without using the built-in function.
11. Write a program to find the maximum element in a set disjoint without using the built-in function.

ANSWER TO MULTIPLE CHOICE QUESTIONS

1.	D	2.	B	3.	D	4.	A	5.	C
6.	C	7.	A	8.	B	9.	C	10.	A
11.	C	12.	A	13.	D	14.	C	15.	C
16.	A	17.	B	18.	D	19.	C	20.	A

8 Functions

LEARNING OBJECTIVES

After studying this chapter, the reader will be able to

- Write and use a user defined function
- Passing parameters to a function
- Write a function with a variable number of arguments
- Return value(s) from a function
- Use namespace
- Write a recursive function
- Write a lambda function with commonly used functions
- Understand generator
- Understand Python modules and user-defined modules
- Understand aliasing, closure and decorators
- Use some special methods and attributes

8.1 INTRODUCTION

There are many situations while writing a program where we need to repeat a particular task many times. Consider a program to calculate the factorial of a number in different places of the program. It would be time-consuming and very inefficient if we need to type the same code for finding factorials every time we need. The solution is to give a name to a piece of code, and every time it is needed, just refer to it.

8.2 FUNCTIONS

When we give a name to a piece of code which does a particular task, it is known as *defining a function*, and every time the piece of code is executed by calling it by its name, it is known as *calling the function*.

There are advantages of using functions. First of all, functions make a program easier to read and understand. Second, functions reduce code duplications. Third, functions allow the code to be reused.

In general, functions appear in a program from three different sources:

- *From Python itself* – Numerous functions (like print()) are an integral part of Python, and are always available without any additional effort; we call these functions *built-in functions*.
- *From Python's preinstalled modules* – A lot of functions, very useful ones, but used significantly less often than built-in ones, are available in a number

DOI: 10.1201/9781003219125-9

of modules installed together with Python; the use of these functions require some additional steps in order to make them fully accessible.

- *Directly from our code* – we can write our own functions, place them inside the code, and use them freely. These functions are known as *user-defined functions.*

The general syntax is:

def Function_Name([parameter list]) :
    body_of_the_function
    [return]

In a function notice that:

- It always starts with the keyword *def* (for "define")
- After *def* comes *name of the function* (the rules for naming functions are exactly the same as for naming variables.)
- Next to the function name, a *pair of parentheses()* is placed. Inside the parenthesis, *input parameters* are written which is optional. Optional means that it is possible to have a function with no input parameters. If more than one parameter is passed, they are separated by a comma.
- The line has to end with a *colon* which is always required.
- The body of the function consists of one or more statements. All those statements inside the body of the function are **tabbed**. Any statement which is not tabbed will not be part of the function.
- At last, there is an optional **return** statement.

Example 8.1 Sample function

```
def First_Function():
  print("My first function")

print("Start...")
print("End...")
```

Output

```
Start...
End...
```

Observe that here is a function with name *First_Function()*. This function contains only print statements. If we run this program, only *Start...* and *End...* will be printed. The print statement *print("My first function")* is not executed. The reason for this is, that even though we have a function but we have never called it. In order to call the function, we need to a code known as *driver code*.

Example 8.2 Sample function

```
def First_Function():
  print("My first function")

print("Start...")
First_Function()
print("End...")
```

Here the driver code consists of a single line, *First_Function()*, which is written below the function definition. It basically means to call the function or invoke it. It executes what is there inside the function definition.

Output

```
Start...
My first function
End...
```

When a function is called, Python remembers the place where it happened and jumps into the invoked function; the body of the function is then executed; once reaching the end of the function forces Python to return to the place directly after the point of invocation. This is shown in figure 8.1.

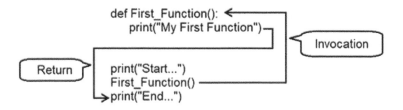

FIGURE 8.1 Calling a function

Do not call a function which is not known at the moment of invocation. As python reads the code from top to bottom. It's not going to look ahead in order to find a function to put in the right place.

Example 8.3 Sample function

```
print("Start...")
First_Function()
print("End...")
def First_Function():
  print("My first function")
```

Observe that the error we are getting is NameError : name "First_Function" is not defined. It means we have not defined the function before calling it.

Output

```
Start...
Traceback (most recent call last):
  File "C:\Users\RakeshN\AppData\Local\Programs\Python\Py
  thon37-32\chap-8.py", line 32, in <module>
  First_Function()
NameError: name "First_Function" is not defined
```

It is advisable to avoid having the same function and variable name.

Example 8.4

```
def First_Function():
  print("My first function")

print("Start...")
First_Function = 1
print("End...")
```

Observe that we have assigned a value to a variable named *First_Function*, it causes Python to forget its previous role of being a function. When we assigned *First_Function = 1*, it treats as a variable so the function named *First_Function()* becomes unavailable. Apart from that while calling a function name, there is always a pair of parentheses to invoke a function.

Output

```
Start...
End...
```

We are free to mix our code with functions. It means, it is not necessary to put all our functions at the top of the source file. But one thing to remember is that *define the function before calling it.*

Example 8.5

```
print("Start...")
  def First_Function():
print("My first function")
First_Function()
print("End...")
```

Output

```
Start...
My first function
End...
```

Once the function is defined it can be called any number of times.

Example 8.6

```
def First_Function():
  print("My first function")
print("Start....")
for i in range(5):
  First_Function()
print("End...")
```

Observe that the function is called five times inside for loop, so as the result, it means *My First Function* is printed five times within the two print statements.

Output

```
Start...
My first function
My first function
My first function
My first function
My first function
End...
```

8.3   PARAMETERIZED FUNCTION

A function can accept data provided by the invoker. Such data can modify the function's behavior, making it more flexible and adaptable to changing conditions. This data is called a *parameter*. As discussed in section 8.2, the parameter is kept inside a pair or parenthesis separated by a comma. A parameter is a variable, but there are two important factors that make parameters different and special from other variables:

- Parameter exists only inside functions in which they have been defined. In other words, it lives inside the function.
- Assigning a value to the parameter is done at the time of the function's invocation, by specifying the corresponding argument. In other words, arguments exist outside functions and are carriers of values passed to corresponding parameters.

Let us take a simple function which is having a single parameter.

Example 8.7 Function with single parameter

```
def Function_with_Parameter(no):
  print("The value passed is: ", no)
print("Function with single parameter ")
Function_with_Parameter(101)
```

Observe that the function definition specifies that the function *Function_with_Parameter()* operates on just one parameter named *"no"*. The value of the argument used during invocation is 101, which has been passed into the function, setting the initial value of the parameter, no.

Output

```
Function with single parameter
The value passed is: 101
```

We must provide as many arguments as there are defined parameters. Failure to do so will cause an error.

Example 8.8 Function with single parameter

```
def Function_with_Parameter(no):
  print("The value passed is: ", no)
print("Function with single parameter ")
Function_with_Parameter()
```

Observe that according to this function definition there is one parameter required, but while calling the function no argument is passed. It means we have not assigned any value to the parameter required for the function. That's why we got a *TypeError* message.

Output

```
Function with single parameter
Traceback (most recent call last):
  File "C:\Users\RakeshN\AppData\Local\Programs\Python\Py
  thon37-32\chap-8.py", line 98, in <module>
  Function_with_Parameter()
TypeError: Function_with_Parameter() missing 1 required
positional argument: 'no'
```

It is possible to have a variable named the same as a function's parameter.

Example 8.9 Function with single parameter

```
def Function_with_Parameter(no):
  no = no+5
  print("The value changed inside the function: ", no)
no=101
print("value before passing", no)
Function_with_Parameter(no)
print("value after passing",no)
```

Observe that the variable name is the same as the parameter and argument. The variable *no* is assigned value 101. This value also passes as an argument. The value before passing the argument and after passing the argument is the same, while inside the function it is added to 5. As a result inside the function, the variable *no* changed to 106. This variable which is inside the function is local to the function. It does not have any effect outside the function. The parameter *no* is a completely different entity from the variable *no* which is outside the function. Changing the parameter's value doesn't propagate outside the function. This also means that a function receives the argument's value, not the argument itself.

Output

```
value before passing 101
The value changed inside the function: 106
value after passing 101
```

8.3.1 POSITIONAL AND KEYWORD PARAMETERS

Assigning the ith argument to the ith parameter is called *positional parameter,* and such arguments are called *positional arguments.* Let us try to understand this with an example.

Example 8.10 Function with positional parameter

```
def Function_with_Parameter(a, b):
  c= a + b
  print("The sum of", a,"and", b, "is" ,c)
Function_with_Parameter(5, 10)
Function_with_Parameter(21, 56)
```

Observe that the same function is called two times with two different arguments. Here the position of the values of the argument is important. When we called the function for the first time the arguments are 5 and 10. These values are passed as parameters. Inside the function parameter, *a* takes the value 5 and parameter *b* takes the value 10.

Output

```
The sum of 5 and 10 is 15
The sum of 21 and 56 is 77
```

Python offers another convention for passing arguments, where the meaning of the argument is dictated by its name, it is called *keyword argument* passing.

Example 8.11 Function with keyword argument passing

```
def Function_with_Parameter(a, b):
  c = a + b
  print("The sum of", a, "and", b, "is", c)
Function_with_Parameter(a=5, b=10)
```

Example 8.12 Function with keyword argument passing

```
def Function_with_Parameter(a, b):
  c = a + b
```

```
   print("The sum of", a, "and", b, "is", c)
Function_with_Parameter(b=10, a=5)
```

Observe that the values passed to the parameters are preceded by the target parameters' names, followed by the = sign, followed by a value. The position doesn't matter here. We mustn't use a non-existent parameter name. Otherwise, we'll get an error message. Each argument's value knows its destination on the basis of the name used. In either case, the result is the same.

Output

```
The sum of 5 and 10 is 15
```

It is allowed to use both positional and keyword parameter methods of argument passing. But the only unbreakable rule is to use positional arguments before keyword arguments.

Example 8.13 Mix fashion parameter passing

```
def Function_with_Parameter(a, b, c):
   sum1 = a + b + c
   print("The sum of", a, ",", b, "and", c, "is", sum1)
Function_with_Parameter(10, 20 ,c=30)
```

Observe the arguments that are passed are *10* and *20* and *c=30*. The first two are positional arguments whereas the third is keyword argument passing.

Example 8.14 Mix fashion parameter passing

```
def Function_with_Parameter(a, b, c):
   sum1 = a + b + c
   print("The sum of" ,a, ",", b, "and", c, "is", sum1)
Function_with_Parameter(10, b=20, c=30)
```

In this example, *a* is the positional argument while *b=20* and *c=30* are keyword arguments.

Example 8.15 Mix fashion parameter passing

```
def Function_with_Parameter(a, b, c):
sum1 = a + b + c
print("The sum of", a, ",", b, "and", c, "is", sum1)

Function_with_Parameter(10, c=30, b=20)
```

In this example, *a* is the positional argument with a value of 10 while *c=30* and *b=20* are keyword arguments. We know the order of keyword arguments is not important.

Output

```
The sum of 10, 20 and 30 is 60
```

8.3.2 DEFAULT PARAMETER

It is also possible to assign a value to the parameter rather than passing a value as an argument. Assume a situation where the value of some argument is always fixed. In that situation we can assign a value to the parameter. The rest can be passed as an argument (may be positional or may be a keyword argument). Such parameters are called *Default Parameters.*

Example 8.15 Assign a value to the parameter

```
def Function_with_Parameter(a, b, c=40):
  sum1 = a + b + c
  print("The sum of" ,a, ",", b, "and", c, "is", sum1)

Function_with_Parameter(10, 20)
Function_with_Parameter(10, b=30)
```

Observe that the function has three parameters. Out of which one parameter is assigned value 40 (i.e., c=40). While calling the function the other two arguments need to be passed (may be positional or may be keyword arguments).

Output

```
The sum of 10, 20 and 40 is 70
The sum of 10, 30 and 40 is 80
```

Even though the default parameters are assigned some values, if we again pass values as arguments for these parameters, the assigned values of the parameters will be overwritten.

Example 8.16 Assign a value to the parameter

```
def Function_with_Parameter(a=1, b=2, c=3):
  sum1 = a + b + c
  print("The sum of" ,a, ",", b, "and" ,c, "is", sum1)

Function_with_Parameter()
Function_with_Parameter(10)
Function_with_Parameter(10, 20)
Function_with_Parameter(10, 20, 30)
```

Observe that the function has default parameter a=1, b=2 and c=3. When we call the function without any argument, it takes the values from the default parameter and prints the result, that is 6.

When we passed only one argument (i.e., 10), It overwrites the value for the default parameter *a*, the values of parameters *b* and *c* are taken from the default parameters (i.e., 2 and 3 respectively). So the output is the sum of 10, 2 and 3, that is 15.

When we passed two arguments as 10 and 20, it overwrites the values of default parameters a and b respectively. The third argument gets its value from the default parameter, which is 3, so the output is the sum of 10, 20 and 3, that is 33.

When we pass all the three arguments as 10, 20 and 30, it overwrites the values of the default parameters a, b and c respectively. So, the output is the sum of 10, 20 and 30, that is 60.

Output

```
The sum of 1, 2 and 3 is 6
The sum of 10, 2 and 3 is 15
The sum of 10, 20 and 3 is 33
The sum of 10, 20 and 30 is 60
```

If we try to pass more than one value for an argument, we'll get a runtime error.

Example 8.17 Error while passing argument

```
def Function_with_Parameter(a, b):
  sum1= a + b
  print("The sum of", a, "and", b, "is", sum1)
Function_with_Parameter(10, a=20)
```

Observe that the first argument is 20. As in positional parameter ordering in which values are passed is important and the first parameter is a, so a takes the value 10. The second argument, a = 20, is a keyword argument. The parameter a is first assigned value 10, and then 20. We get an error message got multiple values for argument 'a'.

Output

```
Traceback (most recent call last):
  File "C:\Users\RakeshN\AppData\Local\Programs\Python\Py
  thon37-32\chap-8.py", line 252, in <module>
    Function_with_Parameter(10,a=20)
TypeError: Function_with_Parameter() got multiple values for
argument 'a'
```

8.3.3 FUNCTION WITH A VARIABLE NUMBER OF ARGUMENTS

We have seen that the number of arguments can never exceed the number of parameters in the definition of the function. Sometimes we don't know how many arguments to pass beforehand. Depending on the situation the number of arguments may vary. In that case, defining a fixed number of parameters is not a good idea.

In order to overcome this, we use *def Function_name(*parameter_name)*. Here *parameter_name* accepts a variable number of arguments. The function call is like any other function call except that we may vary the number of arguments.

Example 8.18 Variable number of arguments

```
def sum_of_numbers(*no):
  sum1 = 0
    for i in no:
    sum1 += i
  return sum1

print("The sum is", sum_of_numbers())
print("The sum is", sum_of_numbers(4))
print("The sum is", sum_of_numbers(34, 2, 52))
lst= [1, 2, 3, 89, 545]
print("The sum is", sum_of_numbers(*lst))
```

Observe the 1st line of the program. *args take a variable number of parameters. The first print statement has 0 arguments, the 2nd print has two arguments and the 3rd print statement has three arguments. The values received by the parameter *agrs are accessed by using a looping structure.

In the 4th line we have defined a list lst=[1, 2, 3, 89, 545]. In the 5th line we passed this list as a parameter. When we pass the list as a parameter, we need to use an asterisk before the list name.

Output

```
The sum is 0
The sum is 4
The sum is 88
The sum is 640
```

The variable length argument can be combined with positional arguments. But the condition is the positional argument must appear before the variable length arguments.

Example 8.19 Variable number of arguments

```
def f1(n1,*var):
  print("Positional Argument", n1)
  print("Variable Length Arguments")
  for i in var:
    print(i)

print("\nwith one argument: ")
f1(10)
print("\nwith two arguments: ")
f1(10,20)
print("\nwith more arguments: ")
f1(34,-3,'Hello',8.2)
```

Observe that the first parameter to the function f1 is a positional parameter and the second parameter is a variable length parameter. In *f1(10)*, the positional argument is 10, and there are no arguments for the variable length parameter. In *f1(10,20)*, the positional argument is 10 and 20 is the argument for the variable length parameter. In *f1(34,-3,'Hello',8.2)*, the positional argument is 34, and the rest of the arguments are for variable length parameters.

Output

```
with one argument:
Positional Argument 10
Variable Length Arguments
with two arguments:
Positional Argument 10
Variable Length Arguments
20
with more arguments:
Positional Argument 34
Variable Length Arguments
-3
Hello
8.2
```

The variable length argument can be combined with default arguments. If we use the default argument as the first parameter, it will always be overwritten by the first argument. If we use the default argument as the last parameter, it will always take the arguments passed for the variable length parameter and it will take the value that is associated with the default parameter, which is evident from the output.

Example 8.20 Variable number of arguments

```
def f1(n1=55,*var):
  print("Positional Argument", n1)
  print("Variable Length Arguments")
  for i in var:
    print(i)
def f2(*var, n1=55):
  print("Positional Argument",n1)
  print("Variable Length Arguments")
  for i in var:
    print(i)

print("Values for f1: ")
print("with one argument: ")
f1(10)
print("with two arguments: ")
f1(10,20)
print("\nValues for f2: ")
```

```
print("with one argument: ")
f2(10)
print("with two arguments: ")
f2(10,20)
```

Output

```
Values for f1:
with one argument:
Positional Argument 10
Variable Length Arguments
with two arguments:
Positional Argument 10
Variable Length Arguments
20
Values for f2:
with one argument:
Positional Argument 55
Variable Length Arguments
10
with two arguments:
Positional Argument 55
Variable Length Arguments
10
20
```

Dictionary can also be passed as an argument to a variable length parameter. We know each item in a dictionary has a key–value pair. In order to access the values for variable length parameter the syntax is: ***def Function_name(**parameter_name)***. Here ****parameter_name** accepts a variable length dictionary. The function call is like any other function call except that we may vary the number of key–value pairs. If we define a dictionary, while passing the dictionary name as a parameter, we need to use two asterisks before the dictionary name.

Example 8.21 Variable number of arguments

```
def Show_Pair(**kwargs):
  for k,v in kwargs.items():
    print(k, "=", v)

print("Key and Value pairs are passed as argument: ")
Show_Pair(Tusarika = 4538562476,Srusti =
57452685421356965,Nakshatra = 58)

print("\nDictionary is passed as argument: ")
d1={"Roll_no":"502","Name":"Sonu","Gender":"Male"}
Show_Pair(**d1)
```

Observe the 2nd line, *for k,v in kwargs.items():*. Here the key and its respective value are received by variables *k* and *v* respectively. In the 5th line, keys and their values are passed separated by an equal (=) sign. In the 7th line, we have defined a dictionary. This dictionary is passed as an argument. But notice the argument in *Show_Pair(**d1)*, two asterisks are used before the dictionary name.

Output

```
Key and Value pairs are passed as argument:
Tusarika = 4538562476
Srusti = 57452685421356965
Nakshatra = 58
Dictionary is passed as argument:
Roll_no = 502
Name = Sonu
Gender = Male
```

8.4   THE RETURN STATEMENT

Functions in any programming language behave similar to their mathematical counterparts where they can produce results. For that, we use the keyword *return*.

When *return* is used inside a function, it causes the immediate termination of the function's execution and an instant return (hence the name) to the point of invocation.

If a function is not intended to produce a result, use of the return statement is not mandatory. It will be executed implicitly at the end of the function.

The return instruction has two different variants.

- The *first* consists of the keyword itself, without anything following it.
- The *second* return variant is extended with an expression.

Example 8.22 Finding factorial

```
def factorial(a):
  if(a<0):
    return
  else:
    fact=1
    for i in range(1, a+1):
      fact= fact * i
    return fact

x= int(input("Find Factorial of "))
res= factorial(x)
print("Factorial of ",x,"is", res)
```

Observe that inside the function there are two return statements. The first return statement appears on the third line and has no expression after it. When the parameter has a value of less than 0, a factorial cannot be calculated; in that case, a return statement without any expression can be used. It will return *None*. None is a very curious value (we have already discussed it in section 2.2.4 Variable Types). Data with value *None*, actually, is not a value at all hence it mustn't take part in any expressions.

None is a keyword. There are only two different circumstances when *None* can be safely used: *First*, when we assign it to a variable. *Second*, while comparing it with a variable.

The second return statement is at the 9th line, and has an expression after the return statement. It will actually return the calculated value as an expression.

Output 1

```
Find Factorial of -9
Factorial of -9 is None
```

Output 2

```
Find Factorial of 6
Factorial of 6 is 720
```

Example 8.23 Centimeter to feet conversion

Figure 8.2 shows that there will be a value for the expression in the return instruction. This statement *transports* the expression's value to the place where the function has been invoked.

In this example, while invoking the function, the argument X = 1, is passed to the parameter *a*. After the calculation, the expression *ft* holds the calculated value 0.03280839895013123 and is returned with the same variable name. That value is passed and *res* holds the value. The result may be freely used here.

Output 1

```
Enter distance in cm: 1
1 cm = 0.03280839895013123 feet
```

Output 2

```
Enter distance in cm: 3
3 cm = 0.09842519685039369 feet
```

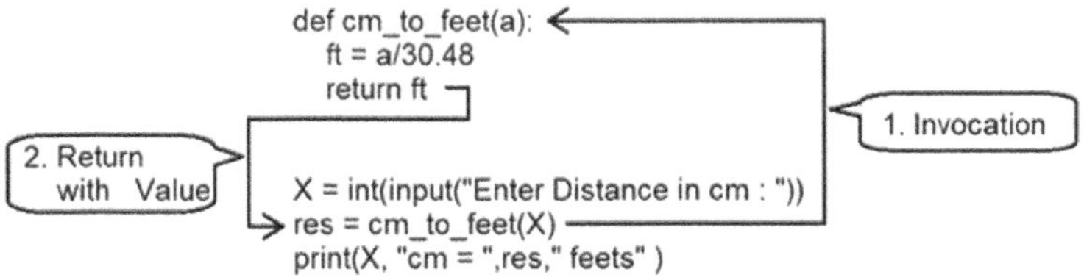

FIGURE 8.2 Explanation of conversion

The returned value may also be completely ignored and lost without a trace. For any return expression, we must have a placeholder. This placeholder will save the value returned by the expression. Without this placeholder, the value returned by the expression is lost.

Example 8.24

```
def Hello():
  print("I said: Hello...")
  return 101
print("Starting...")
Hello()
print("Ending...")
```

Observe that there is a statement *return 101* at line number 3. The value 101 is passed to the main program. But there is no placeholder to accept this value for further processing. So, the value returned is irretrievably lost.

Output

```
Starting...
I said: Hello...
Ending...
```

It is allowed to ignore the function's result, and be satisfied with the function's effect. If a function is indented to return a useful result, it must contain the second variant of the return instruction. It is possible to pass a list as a parameter to a function and it can also return a value or a list.

Example 8.25 List as a parameter, returning a value

```
def List_As_Parametr(lst):
  s=0
  for i in lst:
    s += i
  return s

L= [2,3,6,39,7,9,68,5,54]
sum1= List_As_Parametr(L)
print("The sum of elements in the list is ", sum1)
```

Output

```
The sum of elements in the list is 193
```

Example 8.25 Returning a list

```
def List_As_Parameter(lst):
  L1= []
```

```
   for i in lst:
     if (i % 2 == 0):
        L1.append(i)
   return L1

L1= [2,3,6,39,7,9,68,5,54]
L2= List_As_Parameter(L1)
print("The list with even numbers is ", L2)
```

Observe that list L is passed as an argument, which is a list. This function returned another list with all even numbers for the list which is passed.

Output

```
The list with even numbers is [2, 6, 68, 54]
```

Example 8.26 To find even or odd number

```
def Even_Odd(a):
  if (a % 2 == 0):
    return 'even'
  else:
    return 'odd'

x= int(input("Enter an Integer: "))
res= Even_Odd(x)
print(x,"is",res)
```

Output 1

```
Enter an Integer: 5
5 is odd
```

Output 2

```
Enter an Integer: 88
88 is even
```

8.5   NAMESPACE AND SCOPE OF A VARIABLE

Namespace refers to the space where all the variable names reside. In other words, namespace is a mapping between the variable name and an object. A variable name inside one namespace is different from the variable name outside it. It means a variable name in a namespace is distinct. Layers of namespaces are nothing but the scope. The scope of a variable name is the part of a code, where the variable name is properly recognizable. The scope of a function's parameter is the function itself.

There are three different namespaces in Python. They are: local, global and built-ins.

When the interpreter accesses a variable name for the first time, it looks for it locally. If the variable name is not found in the local namespace, it looks in the global namespace. A global variable uses the keyword *global* before the variable. A global variable can be accessed by any function in the program. The built-in variables are those variables used by Python's built-in functions.

Example 8.27 Scope of a variable

```
def Scope_of_Variable():
  i = 0

Scope_of_Variable()
print(i)
```

Here we have defined a variable *i* inside the function but we are trying to access it outside the function. The result is an error. It is because the variable *i* is defined inside the function, its scope is limited to the function only. So, we cannot access it outside the scope of the variable.

Output

```
Traceback (most recent call last):
  File "C:\Users\RakeshN\AppData\Local\Programs\Python\Py
  thon37-32\chap-8.py", line 475, in <module>
    print(i)
NameError: name 'i' is not defined
```

If the variable is defined outside any function, access it inside the function is allowed.

Example 8.28 Scope of a variable

```
def Scope_of_Variable():
  print("The Variable inside the function", i+1)
i = 10
Scope_of_Variable()
print("The Variable outside the function",i)
```

Observe that *i* = *10* is defined outside the function. We have called the function then we have printed the value of *i* by incrementing the value of *i* inside the function. It can be seen that outside the function the value of *i* remained the same (i.e., *i* = *10*). But inside the function the value of *i* becomes 11. It means a variable existing outside a function has scope inside the functions' bodies.

Output

```
The Variable inside the function 11
The Variable outside the function 10
```

If the same variable name is defined outside as well as inside the function, then the effect of the variable name defined outside is nullified. In other words, the function's variable shadows the variable coming from the outside function.

Example 8.29 scope of a variable

```
def Scope_of_Variable():
  i = 20
  print("The Variable inside the function", i)

i = 10
Scope_of_Variable()
print("The Variable outside the function",i)
```

Observe that variable with same name *i*, is defined outside the function as well as inside the function. The variable inside the function overshadows the variable outside the function. The value of *i* inside the function is 20 while the value of *i* outside the function remains 10.

It means that the scope of a variable existing outside a function is supported only when getting its value. Assigning a value forces the creation of the function's own variable.

Output

```
The Variable inside the function 20
The Variable outside the function 10
```

A function can also be able to modify a variable defined outside. Python has a method which can extend a variable's scope in a way which includes the functions' body. This effect is caused by the keyword *global*. The general syntax is:

global variable_name1, variable_name2,...

Using this keyword inside a function with the variable name(s) separated with commas, forces Python to stop creating new variables inside the function with the same name(s). The variables accessible from outside will be used instead.

Example 8.30 Scope of a variable

```
def Scope_of_Variable():
  global i
  i = 30
  print("The value of i inside the function", i)

i = 10
print("The value of i before function call",i)
Scope_of_Variable()
print("The value of i after function call",i)
```

Observe that inside the function the variable *i* is defined as a global variable. That's why the scope of the variable is extended to the entire program.

Before calling the function the value of *i* is 10, inside the function the value changed to 30. The value of *i*, remained 30 even after the function call.

Output

```
The value of i before function call 10
The value of i inside the function 30
The value of i after function call 30
```

When the argument is a list, changing the value of the corresponding parameter doesn't affect the list.

Example 8.31 Scope of a variable

```
def Scope_of_Variable(lst):
  print("List inside the function", lst)
  lst = [7, 6, 5, 4, 3, 2]

L = [2, 3, 4, 5, 6, 7]
print("The list before function call",L)
Scope_of_Variable(L)
print("The list after function call",L)
```

Observe that even though the value of the parameter we passed is changed inside the function, the list before passing, inside the function and after calling the function is the same. Here we have manipulated the values of the parameter inside the function. As the list inside the function is local to the function, that's why there is no change in the list after calling the function.

Output

```
The list before function call [2, 3, 4, 5, 6, 7]
List inside the function [2, 3, 4, 5, 6, 7]
The list after function call [2, 3, 4, 5, 6, 7]
```

When the argument is a list, changing a list inside the function identified by the parameter will reflect the change outside the function.

Example 8.32 Scope of a variable

```
def Scope_of_Variable(lst):
  print("List inside the function", lst)
  lst.append(50)

L = [2,3,4,5,6,7]
print("The list before function call",L)
Scope_of_Variable(L)
print("The list after function call",L)
```

Observe that inside the function the list is manipulated (3ʳᵈ line, *Lst.append(50)*). Here the update is done by the parameter name. This change to the list reflects outside the function.

Output

```
The list before function call [2, 3, 4, 5, 6, 7]
List inside the function [2, 3, 4, 5, 6, 7]
The list after function call [2, 3, 4, 5, 6, 7, 50]
```

8.6   RECURSIVE FUNCTION

Recursion is a method for solving problems. It involves breaking a problem down into smaller and smaller sub-problems until we get to a small enough problem which can easily be solved.

Recursion is a technique where a function calls itself.

There are three laws of recursion:

- A recursive function must have a base case which tells when the recursive function terminates.
- A recursive function must move toward the base case.
- A recursive function must call itself.

Let us understand these laws by going through some examples.

The factorial has a recursive side too. Generally, we find the factorial of a non-negative Integer as: $n! = 1 \times 2 \times 3 \times ... \times n{-}1 \times n$

It's obvious that: $1 \times 2 \times 3 \times ... \times n{-}1 = (n{-}1)!$

So, we can write the factorial of a as : $n! = (n{-}1)! \times n$

This is in fact a ready recipe for our new solution.

In this example, the base case or the terminating case is factorial of any number is $n{=}=1$ is 1. If the value of n is less than 1 it should return *None*. In order to find the factorial of n we need to find $(n{-}1)! \times n$. For finding $(n{-}1)!$, we need to find $(n{-}2)! \times (n{-}1)$. Continuing like this moving towards the base condition, we need to find $2! = 1! \times 2$. And $1! = 1$.

Example 8.33 Recursive function: factorial

```
def rec_fun_factorial(n):
  if n<1:
    return None
  if n == 1:
    return 1
  else:
    return rec_fun_factorial(n-1)*n

x = int(input("Enter a number: "))
print("The factorial of", x,"is", rec_fun_factorial(x))
```

Observe that the terminating condition is *if the value of 'n' is less than 1*, it returned *None* and *if the value of n is equal to 1* the factorial is 1. Otherwise, we call the function recursively to find the factorial.

Output

```
Enter a number: 6
The factorial of 6 is 720
```

The Fibonacci numbers definition is also a clear example of recursion. The Fibonacci numbers are : 1,1,2,3,...

Fib(a) = Fib(a-1) + Fib(a-2)

They are a sequence of integers built using a very simple rule:

- The first element of the sequence is equal to one.
- The second is also equal to one.
- Every subsequent number is the sum of the two preceding numbers Fib(a) = Fib(a-1) + Fib(a-2)

Here are first few Fibonacci numbers,

Fib1 = 1
Fib2 = 1
Fib3 = 1 + 1 = 2
Fib4 = 1 + 2 = 3
Fib5 = 2 + 3 = 5

Example 8.33 Recursive function: Fibonacci number

```
def rec_fun_fibo(a):
  if a<1:
    return None
  if a<2:
    return 1
  else:
    return rec_fun_fibo(a-1) + rec_fun_fibo(a-2)

x= int(input("Enter a number: "))
print("The", x,"fibonaccii number is", rec_fun_fibo(x))
```

Output

```
Enter a number: 9
The 9 fibonaccii number is 34
```

8.7   LAMBDA FUNCTION

A lambda function is a function without a name (we can also call it an anonymous function). In many cases, the lambda function can exist and work while remaining fully incognito. Lambda functions are not declared as other functions using the *def* keyword. Rather, they are created using the *lambda* keyword. Lambda functions are throw-away functions, i.e., they are just needed where they have been created and can be used anywhere a function is required. The lambda feature was added to Python due to the demand from LISP programmers.

The general syntax of the lambda function is:

lambda parameters : expression

In a set notice that:

- A lambda function *starts* with a keyword *lambda*.
- Next to the keyword the *parameters* are written, which is optional.
- It is followed by *colon*.
- The *expression* is written that need to be evaluated after the colon.
- The lambda function *returns* a value.

Example 8.34

```
First = lambda : 2 + 3
print("The sum is ", First())
```

Here the 1st line is the lambda function which returns the sum of 2 and 3. After the keyword lambda and the colon, the expression 2 + 3 is written. The calculated value of this expression is returned. This value is received by a variable named *First*. This variable name is used to invoke the lambda function *First()*. We can say that the function is not anonymous anymore. It does not take any parameters but it returns the sum.

Output

```
The sum is 5
```

Example 8.35

```
Second = lambda x: 2 * x
print("The multiplication is ", Second(5))
```

Observe that in this lambda function, after the keyword lambda, a parameter *x* is written. The expression of the function *2 * x* is written after the colon. This function returns the value of 2 * argument. This lambda function is invoked in the 2nd line as *Second(5)*.

Output

```
The multiplication is 10
```

Example 8.36

```
First = lambda : 3
Third = lambda y, z : y**z
print("The power is ", Third(5,First()))
```

Observe that there are two lambda functions that are defined at the 1st and 2nd line. The lambda function in the 1st line does not take any parameter and returns 3. The second lambda function defined in the 2nd line takes two parameters and returns the value *(y**z)*. In the 3rd line, we have invoked the function with *Third(5,First())*. Observe that the second parameter is the returned value of the first lambda function, which returns 3. As a result, *Third(5,First())* calculated *Third(5,3)*. Which returned 5**3, that is 125.

Output

```
The power is 125
```

The most interesting part of using lambdas is, that it appears when we can use them in their pure form – as anonymous parts of code intended to evaluate a result.

The main advantages are:

- In the lambda function the code becomes *shorter, clearer* and more *legible*.
- The lambda function internally *returns* the expression value and we are not required to write a return statement explicitly.
- Sometimes we can pass a function as an argument to another function. In such cases, lambda functions are the best choice.

We can use lambda functions very commonly with inbuilt functions like *filter()*, *map()* and *reduce()*.

8.7.1 FILTER()

This function filters its second argument while being guided by directions flowing from the function specified as the first argument. The values for which the lambda function returns True, pass the filter and other values are rejected.

Example 8.36 filter() function

```
from random import seed,randint
seed()
My_lst= [randint(-20,20) for x in range(10)]
filtered= list(filter(lambda x: x>0 and x%2==0, My_lst))
print("List before filter...", My_lst)
print("List after filter...", filtered)
```

Observe that we've made use of the random module. To initialize the random number generator with the *seed()* function, and to produce 10 random integers from –20 to 20 using the *randint()* function. The list is then filtered, and only the numbers which are even and greater than zero are accepted. Observe the lambda function defined in 4th line, *lambda x:x>0 and x%2==0, My_lst*. It takes all the values from *My_lst*. It checks *if the values are greater than 0 and even number.* Those values which satisfy these conditions are filtered and are written to another list *filtered.*

Output

```
List before filter... [3, 9, 1, -9, 5, -17, 8, -11, 3, 9]
List after filter... [8]
```

8.7.2 MAP()

This function is used to apply some functionality and generate new elements with the required modification to a given sequence. The general syntax is: ***map(function,list)***.

The *map()* function takes two arguments, a function and a list. The *map()* function applies the function passed by its first argument to all the elements of its second argument. It returns an iterator. We can use the resulting iterator in a loop or convert it into a list.

Example 8.37

```
from random import seed, randint
seed()
My_lst= [randint(-20,20) for x in range(5)]
mapped1= list(map(lambda x: x>0 and 2*x, My_lst))
mapped2= list(map(lambda x: 2*x , My_lst))
print("List before map...", My_lst)
print("Mapped List with some element...", mapped1)
print("Mapped List with all element...", mapped2)
```

Observe 4th and 5th lines of the program. In the 4th line, a lambda function is used to generate five positive random numbers. When we applied *map()* to this result, it returned *False,* for all those elements not qualifying the lambda function. Here the negative numbers in *My_lst* does not qualify the part of the expression *x>0 and 2 * x,* so in that place the output is *False.*

In the 5th line, the lambda function generates items *multiplied by 2* to all the items in *My_lst.* The *map()* function is applied to all these items and it qualifies the function *2*x,* so it prints all the elements.

Output

```
List before map... [8, -17, -17, 17, -15]
Mapped List with some element... [16, False, False, 34, False]
Mapped List with all element... [16, -34, -34, 34, -30]
```

8.7.3 REDUCE()

The *reduce()* function takes two arguments as in *map()*.The function is called with the first two items from the sequence and the result is generated. Again, it is called with the result obtained in the previous step and the next value in the sequence. This process keeps repeating until there are no more items in the sequence. This function is used to apply to all the elements in a sequence. This function is defined in the *functools* module (see example 3.38). The general syntax is: *reduce(function,list)*.

Example 8.38

```
from functools import*
from random import seed, randint
seed()
My_lst= [randint(-20,20) for x in range(5)]
reduced1= reduce(lambda x,y: x + y, My_lst)
print("List before filter...", My_lst)
print("Reduced list with some element...", reduced1)
```

Observe the 5th line of the program. A lambda function is defined that takes two arguments and returns their sum. The *reduce()* function is applied to it. The reduce function returns the result as –10.

Output

```
List before filter... [-19, 12, 7, 10, -20]
Reduced list with some element... -10
```

8.7.4 ZIP()

This function is an aggregator. It combines two or more sequences. The individual items are combined to form a tuple. The *zip()* functions exits at the end of the smallest list. The general syntax is: *zip(iterable 1, iterable 2, …, iterable n)*.

Example 8.39

```
tup1= [1, 2, 3]
tup2= ['Peacock', 1.5, 'Owl', 100, 2+7j]
print("The first tuple is", tup1)
print("The second tuple is", tup2)
new_tup= tuple(zip(tup1,tup2))
print("The new tuple after zip() is ", new_tup)
```

Observe that the length of tup1 is 3 and the length of tup2 is 5. After applying *zip()*, the resultant tuple has a length 3. Each element in the *new_tup* is a pair, where the first element is from tup1 and the second element is from tup2.

Output

```
The first tuple is [1, 2, 3]
The second tuple is ['Peacock', 1.5, 'Owl', 100, (2+7j)]
The new tuple after zip() is ((1, 'Peacock'), (2, 1.5),
  (3, 'Owl'))
```

8.8 GENERATORS

A generator is a function which is responsible for generating a sequence of values. We can write generator functions just like ordinary functions, but it uses the *yield* keyword to return values. Iterators are used to fetch one value at a time. For that we need to use two methods *iter()* and *__next__()* or *next(iter)*. The syntax of the generator is the same as a function, but instead of the *return* statement, a *yield* statement is executed.

In order to use the value given by *yield,* we need to use *__next__()* or *next(iter)*. In any function we can have at most one return statement, where as in a generator we can have *at least one yield statement*. The return statement terminates the function but the yield statement does not terminate the generator. The most convenient way for creating iterators in Python is through the use of generators.

Example 8.40

```
from random import *
def anynumber():
  yield random()
i = anynumber()
print("The random number is", i.__next__())
i = anynumber()
print("The random number is", next(i))
```

Observe that there is a random number generated by using the *random()* function. The generated value is returned by a *yield*. In order to access this value, we need to use an iterator *__next__()*.

Output

```
The random number is 0.27933202520236555
The random number is 0.08214523316470157
```

It is possible to have more than one yield statement in a generator. There are two different ways to access the yield statements. The first method is by writing separate print statements and accessing each yield by *next(iter)* or *iter.__next__()*. The second method is by using a looping structure.

Example 8.41

```
from random import *
def anynumber():
```

```
  yield random()
  yield randint(100,110)
j = anynumber()
print("First method...")
print("The random number is", next(j))
print("The random Integer is", j.__next__())

print("Second method...")
for i in anynumber():
  print("The random number is", i)
```

Observe that in the generator *anynumber()*, there are two yield statements. In the first method of accessing the values, we have used an iterator *j = anynumber()*, and for each yield statement a separate print statement is written by using either *next(j)* or *j.__next__()*. In the second method, we have used a looping structure, each yield statement is accessed by the iterator of the looping structure with a single print statement inside the loop.

Output

```
First method...
The random number is 0.12800752115346337
The random Integer is 105
Second method...
The random number is 0.34107337461558385
The random number is 102
```

Sometims yield generates values in the form of a list. In that case we can access these values using a looping structure.

Example 8.42

```
def even(n):
  i = 1
  while (i <= n):
    if (i%2==0):
      yield i
    i += 1
j = even(10)
for k in j:
  print(k)
```

Observe that we are using a generator *even()*, It yields an even number below 10. Here j is the list which holds all the values yielded by this generator. Like any other list items, we can print them by using a looping structure.

Output

```
2
4
```

```
6
8
10
```

Sometimes when we are dealing with a large number of data, it is always advisable to use a generator rather than a function. This is because a function returns a large amount of data at a time and all of them will be loaded into the memory (sometimes the memory may not be sufficient even.). And we may be working with only one data at a time. Under such situations we can use a generator which yields only one data at a time. By using a generator, instead of fetching the entire data, we can fetch one data at a time.

8.9 PYTHON MODULES

Assume that we have written some functions that we intend to use later in some other program. We have two different ways to do this. *First*, we can cut and paste the source code of the function in the new program, which would be very time-consuming and inefficient. The code will face a growing problem. A larger code always means tougher maintenance. Searching for bugs is always easier when the code is smaller. Moreover, when the code being created is expected to be really big, we may want to divide it into many parts, implemented in parallel by a few individual developers. Of course, this cannot be done using one large source file, which is edited by all programmers at the same time. This will surely lead to a disaster. If we want such a software project to be completed successfully, we have to have the means allowing us to:

- Divide all the tasks among the developers.
- Join all the created parts into one working whole.

The process of dividing is often called *decomposition*. To address this issue we have module.

Second, we can save the function in a file with dot py (.py) extension and use the import statement to invoke that function. Python has a way to put definitions in a file and use them in a script or in an interactive instance of the interpreter. Such a file is called a *module*.

The handling of modules consists of two different issues:

- To create a brand-new module.
- To use an already existing module.

A module is identified by its name. If you want to use any module, we need to know the name. A number of modules are delivered together with Python itself. All these modules, along with the built-in functions, form the *Python Standard Library*. If we want to take a look at the full list of all *volumes* collected in that library, we can find it at https://docs.python.org/3/library/index.html.

Each module consists of *entities*. These entities can be functions, variables, constants, classes and objects. If we know how to access a particular module, we can make use of any of the stored entities. Python includes many modules, for example, there is a module called *math*. This module includes lots of mathematical functions, such as sine, cosine, square root, log and many more.

Example 8.43

```
print(sqrt(100))
```

Output

```
Traceback (most recent call last):
  File "C:\Users\RakeshN\AppData\Local\Programs\Python\Py
  thon37-32\chap-8.py", line 882, in <module>
    print(sqrt(100))
NameError: name 'sqrt' is not defined
```

Observe that we got an error *name 'sqrt' not defined*. It means, that we have neither explicitly defined a function nor have we used any module in which the *sqrt()* function is defined. Before using the *sqrt()* function, which is defined in the module math, we must import it.

8.9.1 IMPORTING A MODULE

To make a module usable, we must import it. Importing a module is done by an instruction named with the keyword *import*.

There are *three* different ways of importing a module to use the functions defined in it.

1. The first and the simplest way to import a particular module is to use the import instruction as ***import Module_Name***.

The clause contains:

- The *import* keyword.
- The *name of the module* which is subject to import.

The instruction may be located anywhere in our code, but it must be placed before the first use of any function defined in the module.

If we want to import more than one module, we can do it by repeating the import clause, or by listing the modules separated by a comma after the import keyword. For example: ***Import math, sys***

If the module of a specified name exists and is accessible, Python imports its contents, but they don't enter the code's namespace. This means that we can have our own entities named *sqrt()* or *pi* and they won't be affected by the import in any way. In order to access the *pi* coming from the math module, we have to qualify *pi* with the name of its original module. To qualify the names of *pi* and *sqrt* with the name of its originating module *math.pi* or *math.sqrt*

The clause contains:

- The name of the module, which is math here.
- A dot.
- The name of the entity, which is pi or sqrt here.

Such a form clearly indicates the namespace in which the name exists.

Example 8.44

```
import math
print("Square root of 100 is", math.sqrt(100))
```

The first statement is the import statement. In order to use the built-in function *sqrt()*, which is defined inside the math module, we must write *math.sqrt()*.

Output

```
Square root of 100 is 10.0
```

Using this qualification is compulsory if a module has been imported by the import module instruction. It doesn't matter if any of the names from the code and from the module's namespace are in conflict or not.

Example 8.45

```
import math
pi= 3.14
print("The value of pi is", pi)
print("The value of pi in module math is", math.pi)
```

Observe that we have assigned *pi* to 3.14. Again, *pi* is a function defined in the *math* module. In order to use the built-in function, we need to qualify it by writing *math.pi*. Of course, we are having a variable name *pi*, there is no conflict while using them as we have a mechanism of qualification.

Output

```
The value of pi is 3.14
The value of pi in module math is 3.141592653589793
```

2. The second method for importing modules is to precisely point out which functions of the module are to be imported. If more than one function is to be imported from the same module, separate them by a comma.

The import instruction is as follows: *from Module_Name import Function_Name*. The instruction consists of the following elements:

- The *from* is a keyword.
- The *name of the module* to be imported.
- The *import* is a keyword.
- The *name or list of names* of the function/functions which are being imported.

The instruction has this effect:

- The listed functions only are imported from the indicated module.
- The names of the imported entities are accessible without qualification.

Example 8.46

```
from math import sqrt
x = 100
print("The square root of",x,"is", sqrt(x))
```

Observe that the way we have used the import statement. In this way, we have imported a specific function i.e., *sqrt()* from the *math* module. And we have called the function in the print statement without qualification. But if we have many functions that we want to use, we have to explicitly type the name of these functions on the import statement separated by a comma, which is difficult.

Output

```
The square root of 100 is 10.0
```

3. In the third method, the import's syntax is more aggressive. The import instruction is as follows: *from Module_Name import ***.

Observe that the name of a function (or the list of functions' names) is replaced with a *single asterisk* (*). With this format, all the functions which are included in the module will be automatically imported.

It is unsafe to use this method, unless we know all the names provided by the module, we may not be able to avoid name conflicts.

Example 8.47

```
from math import *
x=100
print("The Square root of", x, "is", sqrt(x))
print("Sin 90 degree", sin(pi/2))
```

Observe the import statement, we have used *from math import* * . In the program we have used three functions defined in the math module that is *sqrt()*, *sin()* and *pi* without qualification.

Output

```
The Square root of 100 is 10.0
Sin 90 degree 1.0
```

8.9.2 ALIASING

If we use the first or second method of importing a module i.e., *import Module_ Name* or *from Module_Name import Function_Name*, and we don't like a particular module's name or function's name, we can give it any name we like. This is called *aliasing*. Aliasing causes the module or function to be identified under a different name than the original. This may shorten the qualified names. Aliasing is done together with importing the module.

The syntax for creating an alias is:

Import Module_Name as alias
    or
Form Module_Name import Function_Name as alias

In the first one, *Module_Name* identifies the original name of the module while the *alias* is the name we wish to use instead of the original.

In the second one, *Function_Name* identifies the original name of the function which is defined in the Module while the *alias* is the name we wish to use instead of the original.

Example 8.48

```
import math as M
x= 100
print("The square root of",x,"is",M.sqrt(x))
print("Sin 90 degree= ", M.sin(M.pi/2))
```

Observe that we have used an alias *M* for the module math. While calling the function we have used as *M* for qualification.

Output

```
The square root of 100 is 10.0
Sin 90 degree= 1.0
```

Example 8.49

```
from math import sqrt as SQ, pi as PI, sin as S
x= 100
```

```
print("The square root of", x,"is", SQ(x))
print("Sin 90 degree= ", S(PI/2))
```

Observe that we have used an alias for function names in this example also. We have used aliases SQ for *sqrt()*, PI for *pi* and S for *sin()*. While calling the function the alias names are used.

Output

```
The square root of 100 is 10.0
Sin 90 degree= 1.0
```

After successful execution of an aliased import, the original module name becomes inaccessible and must not be used.

8.9.3   USER-DEFINED MODULE

We can also create our own module. Writing our own modules doesn't differ much from writing ordinary programs. We need two files. The first one is the module itself. The second file contains the code using the new module. The first function adds two numbers and returns their sum. The second function multiplies two numbers and returns the result. We are going to save these two functions in a file called *my_module.py*.

Example 8.50 my_module.py

```
def MY_add(a,b):
  return a + b

def My_mul(a,b):
  return a * b
```

Now open another file to use the function defined in *my_module.py* as described in example 8.50. We can invoke the functions defined in *my_module.py* by using any of the methods discussed earlier.

Example 8.51

```
import my_module
a= 5
b= 10
x= my_module.MY_add(a,b)
print("The sum is",x)
y= my_module.My_mul(a,b)
print("The multiplication is", y)
```

Observe that we have used the function *MY_add()* and *My_mul()* which we have defined in our previous program. But before we call the functions, we must import them. The first line of the program *import my_modul* tells Python that the functions *MY_add()* and *My_mul()* functions are stored in *my_module*. While calling the function, we write *my_modul.MY_add()* and *my_modul.My_mul()* (i.e., module_Name. Function_Name.)

In the second method, the import statement is in the form *from Module_Name import function_Name*. If more than one function is used separate them by a comma.

Example 8.52

```
from my_module import MY_add,My_mul
a=5
b=10
x= MY_add(a,b)
print("The sum is",x)
y= My_mul(a,b)
print("The multiplication is", y)
```

Observe, in the first line we have written: *from my_modul import MY_add, My_mul*. While calling the function, we call the function name only. But if we have many functions that we want to use, we have to explicitly type the name of these functions on the import statement, which is difficult. In the third method, we can import all the functions which are included in a file *my_modul.py* and use as per our needs. When we write *from module_name import *, with this format, all the functions which are included in the module, will be automatically imported.

Example 8.53

```
from my_module import *
a=5
b=10
x= MY_add(a,b)
print("The sum is",x)
y= My_mul(a,b)
print("The multiplication is", y)
```

Output

```
The sum is 15
The multiplication is 50
```

8.10   CLOSURES

In a nested function call, when the objects of an outer function are available in an inner function, it is called *scope of the variable*.

Example 8.54

```
def Outer_Fun(n):
  print("the value inside the outer function is", n)

    def Inner_Fun():
      m = n + 1
      print("the value inside the inner function is", m)

  Inner_Fun()

Outer_Fun(5)
```

The argument passed to the *Outer_Fun()* is 5. As the scope of the parameter is inside the entire scope of the *Outer_Fun()*, it is available inside the *Outer_Fun()* as well as inside the *Inner_Fun()*. In order to invoke the *Inner_Fun()*, we need to call the function somewhere, after defining it, inside the *Outer_Fun()*.

Output

```
the value inside the outer function is 5
the value inside the inner function is 6
```

But if we want to access the variable defined inside *Inner_Fun()*, m, we get an *NameError*.

Example 8.55

```
def Outer_Fun(n):
  print("the value inside the outer function is", n)

  def Inner_Fun():
    m = n + 1
    print("the value inside the inner function is", m)

    Inner_Fun()
    print("the value outside the inner function but inside
    the outer function is", m)

Outer_Fun(5)
```

The error is obvious because the variable m, which is defined inside the *Inner_Fun()*, is accessed outside its scope.

Output

```
the value inside the outer function is 5
the value inside the inner function is 6
Traceback (most recent call last):
```

```
File "C:\Users\RakeshN\AppData\Local\Programs\Python\Py
thon37-32\chap-8.py", line 1005, in <module>
  Outer_Fun(5)
File "C:\Users\RakeshN\AppData\Local\Programs\Python\Py
thon37-32\chap-8.py", line 1003, in Outer_Fun
print("the value outside the inner function but inside the
outer function is", m)
NameError: name 'm' is not defined
```

Let us make some changes to the programs defined in example 8.55.

Example 8.56

```
def Outer_Fun(n):
  print("the value inside the outer function is ", n)

  def Inner_Fun():
    m = n + 1
    print("the value inside the inner function is", m)

  return Inner_Fun
Another_Fun = Outer_Fun(5)
Another_Fun()
```

Here observe the last line of *Outer_Fun(n)*, Instead of calling the *Inner_Fun()*, we have written *return Inner_Fun*. As we are returning a function, we need to bind it to a function, *Another_Fun*. This can be invoked by *Another_Fun()*.

Observe the output, by invoking the *Outer_Fun()* function, we could access the data in the inner function.

Output

```
the value inside the outer function is 5
the value inside the inner function is 6
```

- The *Inner_Fun()* function returns the value of the variable accessible inside its scope, as *Inner_Fun()* can use any of the entities at the disposal of *Outer_Fun()*.
- The *Outer_Fun()* function returns the *Inner_Fun()* function itself.
- This technique of attaching the data in the *Inner_Fun()* function to the *Outer_Fun()* is called *closure*.
- It returns a *copy* of the *Inner_Fun()* function.
- The *Inner_Fun()* was frozen at the moment of *Outer_Fun()*'s invocation.
- The frozen function contains the state of all local variables, which also means that the value of *m* is successfully retained, although *Outer_Fun()* ceased to exist.

The conditions under which the concept of closure works is:

- A nested function must be used.
- The outer function must refer to a value inside the inner function.
- The inner function must be returned.

The concept of closure provides data hiding to some extent.

Example 8.57

```
def mkpower(par):
  expo = par
    def pow(p):
       return p**expo
  return pow
sq= mkpower(2)
cu= mkpower(3)

for i in range(2,5):
  print(i,"has Square", sq(i), "and Cube", cu(i))
```

Output

```
2 has Square 4 and Cube 8
3 has Square 9 and Cube 27
4 has Square 16 and Cube 64
```

8.11   DECORATORS

Most of the time we use built-in functions and at times we define a function inside a user-defined module. These functions are general-purpose functions. It means mostly we use these functions as it is. When we want some additional feature to these existing functions, we cannot do it because the code of the function is not available to us. In order to provide this feature, we have the *decorator*. Decorator gives additional meaning to the existing functions. In other words, decorators *wraps* a function with another function. Decorators can also be applied to user-defined functions.

Decorator is a function which can take a function as an argument and extend its functionality and return a modified function with extended functionality. While extending the functionality, the behavior of the original function does not change.

Example 8.58

```
from My_TestModule import Div
n,m= eval(input("Enter two numbers (Separated by a coma): "))
print("The division is", Div(n,m))
```

Observe that a function *Div(n,m)* is called, which is defined in *My_TestModule*. Only what we know is how to use the function, we don't have any knowledge about the code of the function.

This function returns the quotient when *a is divided by b*. When a = 24 and b = 6, the result is 4.0.

Output

```
Enter two numbers (Separated by a coma): 24,6
The division is 4.0
```

When a = 6 and b = 24, the result is 0.25

Output

```
Enter two numbers (Separated by a coma): 6,24
The division is 0.25
```

When a = 33 and b = 0, it raised an error message.

Output

```
Enter two numbers (Separated by a coma): 33,0
Traceback (most recent call last):
  File "C:\Users\RakeshN\AppData\Local\Programs\Python\Py
  thon37-32\chap-8.py", line 1079, in <module>
  print("The division is", Div(n,m))
  File "C:\Users\RakeshN\AppData\Local\Programs\Python\Py
  thon37-32\My_TestModule.py", line 2, in Div
  return a/b
ZeroDivisionError: division by zero
```

Now we want to add some functionality to this function. In this division, we want to make the dividend always greater than the divisor. If the dividend is less than the divisor we will make the division by swapping the divisor and the dividend. And it should return 0 if either divisor or the dividend (but not both) is 0. As the code is not available with us, we will use a decorator to add the functionality.

Example 8.59

```
import My_TestModule

def improved_Div(func):
  def inner(a,b):
    if a < b and a == 0:
      return func(0,b)
    elif b == 0:
      return func(0,a)
    elif a < b :
      return func(b,a)
    else:
          return func(a,b)
  return inner
```

```
n,m= eval(input("Enter two numbers (Separated by a coma): "))
Div = improved_Div(My_TestModule.Div)
print("The division is", Div(n,m))
```

Output

```
Enter two numbers (Separated by a coma): 3,6
The division is 0.5
```

Output

```
Enter two numbers (Separated by a coma): 6,3
The division is 2.0
```

Output

```
Enter two numbers (Separated by a coma): 0,6
The division is 0.0
```

Output

```
Enter two numbers (Separated by a coma): 6,0
The division is 0.0
```

Output

```
Enter two numbers (Separated by a coma): 0,0
Traceback (most recent call last):
  File "C:\Users\RakeshN\AppData\Local\Programs\Python\Py
  thon37-32\chap-8.py", line 1100, in <module>
    print("The division is", Div(n,m))
  File "C:\Users\RakeshN\AppData\Local\Programs\Python\Py
  thon37-32\chap-8.py", line 1091, in inner
    return func(0,a)
  File "C:\Users\RakeshN\AppData\Local\Programs\Python\Py
  thon37-32\My_TestModule.py", line 2, in Div
    return a/b
ZeroDivisionError: division by zero
```

Until now we have used decorators for built-in functions. Decorators can be used on a user-defined function in the same manner as we have used them for built-in functions with little modification.

Example 8.60

```
def improved_Div(func):
  def inner(a,b):
    if a < b and a==0:
      return func(0,b)
    elif b == 0:
      return func(0,a)
```

```
      elif a < b:
        return func(b,a)
      else:
        return func(a,b)
      return inner

  @improved_Div
  def Div(a,b):
    return a/b

  n,m= eval(input("Enter two numbers(seperated by comma):"))
  print("The division is", Div(n,m))
```

Observe that we have a user-defined function *Div(a,b)*. This function is wrapped up by another function *improved_Div(func)*. While invoking, we call *Div(n,m)*, which automatically refers to *improved_Div(func)* as *@improved_Div"* written before *"def Div(n.m)*. The output remains the same as in example 8.62. Remember that the wrapping function should be defined before the wrapped function.

8.11.1   PASSING A PARAMETER TO A DECORATOR

We know that in order to perform some tasks, sometimes we need to pass arguments. As in earlier examples, we passed two integers to perform Div() operation. Similarly, we can also pass parameters to the wrapping function.

Example 8.61

```
def improved_XYZ(x,y,msg):
  def inner1(func):
    def inner2():
      if x <= y :
        print(msg,"-ADD-",x + y)
      else:
        print(msg,"-SUB-", x - y)
    return inner2
  return inner1

@improved_XYZ(x= int(input('x = ')),y = int(input('y = ')),
msg = "Hello")
def XYZ():
  pass

XYZ()
```

Observe that we are having a function *XYZ()* which practically does not perform anything. The decorator is *@improved_XYZ()*, it took three parameters. We can take positional parameters or keyword parameters.

Output

```
x = 4
y = 8
Hello -ADD- 12
```

Output

```
x = 8
y = 4
Hello -SUB- 4
```

8.11.2   DECORATOR CHAINING

Multiple decorators can also be applied to a function. This is also called decorator chaining or decorator stacking. The decorators will be applied, starting with the inner decorator.

Example 8.62

```
def improved_Div2(func):
  def inner(a,b):
    if a < b:
      return func(b,a)
    else:
      return func(a,b)
  return inner

def improved_Div1(func):
  def inner(a,b):
    if a < b and a == 0:
      return func(0,b)
    elif a > b and b == 0:
      return func(0,a)
    else:
      return func(a,b)
  return inner

@improved_Div2
@improved_Div1
def Div(a,b):
  return a/b

n,m= eval(input("Enter two numbers(seperated by comma):"))
print("The division is", Div(n,m))
```

Here we have modified example 8.60. Here we created two decorators @ *improved_Div1*, the inner decorator and @*improved_Div2*, the outer decorator.

The inner decorator is called first then the outer decorator. The output is the same as example 8.60.

Example 8.63

```
def decor2(func):
  def inner():
    x= func()
    print("Decorator 2", 4*x)
    return 4*x
  return inner

def decor1(func):
  def inner():
    x= func()
    print("Decorator 1", x-2)
    return x-2
  return inner

@decor2
@decor1
def XYZ():
  no = int(input("Enter a number: "))
  return no

print(XYZ())
```

Output

```
Enter a number: 3
Decorator 1 1
Decorator 2 4
4
```

8.12   SOME SPECIAL METHODS AND ATTRIBUTES

8.12.1   _ _ NAME _ _

It is a special attribute that returns the name of the module. If the source file is executed as the main program, it returns __main__ and if the file is imported from any module, it returns the module name. The general syntax is: ___main___.

Example 8.64

```
print(__name__)
import math
print(math.__name__)
```

Output

```
__main__
math
```

8.12.2   _ _ MODULE _ _

It is a special attribute that returns the module to which the class or method belongs.
The general syntax is: ___module___.

Example 8.65

```
from math import sin
print(sin.__module__)
```

Output

```
math
```

WORKED OUT EXAMPLES

Example 8.66 Write a recursive function that converts decimal to octal

```
def Dec_To_Oct(n):
  if n > 1:
    Dec_To_Oct(n//8)
  print(n % 8 , end = '')

# decimal number
dec = int(input("Enter a Decimal number "))
Dec_To_Oct(dec)
print(" is octal of ", dec)
```

Output

```
Enter a Decimal number 10
12 is octal of 10
```

Example 8.67 Write a program that displays a calendar of a given month and year

```
import calendar
yy = int(input("Enter Year: "))
mm = int(input("Enter Month: "))
print(calendar.month(yy,mm))
```

Output

```
Enter Year: 2021
Enter Month: 04
    April 2021
Mo Tu We Th Fr Sa Su
            1 2 3 4
 5  6  7  8  9 10 11
12 13 14 15 16 17 18
19 20 21 22 23 24 25
26 27 28 29 30
```

Example 8.68 Write a program that checks whether a String is a palindrome or not

```
def palin(my_str):
  my_str = my_str.casefold()
  rev_str = reversed(my_str)
  if list(my_str) == list(rev_str):
    print(my_str, "The String is a Palindrome")
  else:
    print(my_str,"The String is not a Palindrome")

str1 = input("Enter a String ")
palin(str1)
```

Output

```
Enter a String Madam
madam The String is a Palindrome
```

MULTIPLE CHOICE QUESTIONS

1. Consider the following code
   def get_score(total = 0, valid = 0):
       result = int(valid)/int(total)
       return result

   Which statement is incorrect?
   a. score=get_score(40,4)
   b. score=get_score('40','4')
   c. score=get_score(40)
   d. score=get_score(0,10)

2. Consider the following code
   data = []
   def get_data():
       for i in range(1,5):
           marks = input("Enter Marks")

```
        data.append(marks)
    def get_avg():
        sum = 0
        for mark in data:
            sum += mark
            return sum / len(data)
    get_data()

    print(get_avg())
```

For the input: 20,30,40,50 what is the result?

a. 5.0
b. 5
c. NameError
d. TypeError

3. Consider the following code

```
    def get_names():
        names = ['Parrot','Owl','Peacock','Maina','Crow']
        return names[2:]
    def update_names(elements):
        new_names = []
        for name in elements:
            new_names.append(name[:3].upper())
            return new_names

    print(update_names(get_names()))
```

Find the output.
a. ['PEA']
b. ['PAR','PEA']
c. ['PAR','PEA','OWL']
d. ['PAR','PEA','OWL','MAI']

4. Consider the following code

```
    def my_list( # Your Code):
        lst.append(a)
        return lst
    my_list('Peacock')
    my_list('Parrot')
    print(my_list('Pigeon'))
```

Which statement should be inserted in place of xx to get the output:
['Peacock', 'Parrot', 'Pigeon']

a. a,lst=()
b. a,lst=[]
c. a,lst={}
d. a,lst=0

5. Consider the following code

```
def f1(x = 0, y = 0):
    return x+y
```

Which among the following is incorrect?
a. f1()
b. f1(10)
c. f1('10', '20')
d. f1('10')

6. Consider the following code

```
def calculate(amount = 6, factor = 3):
    if amount > 6 :
        return amount * factor
    else:
        return amount * factor *2
```

Which among the following statement will give the output 30?
a. calculate()
b. calculate(10)
c. calculate(5,6)
d. calculate(6,3)

7. Find the output of the following code

```
def f1(a,b):
    return a+b
print(f1(b=1,2))
```

a. 1
b. 2
c. 3
d. Error

8. Find the output of the following code

```
st = 'xyzt'
def f1(s):
    del s[2]
    return s
print(f1(st))
```

a. 'yzt'
b. 'xzt'
c. 'xyt'
d. Error

9. Find the output of the following code
```
def f1(x):
    return 1 if x%2 !=0 else 2
print(f1(f1(1)))
```

   a. 1
   b. 2
   c. None
   d. Error

10. Which of the following is a proper invocation?
```
def fun(a,b,c=0)
```
   a. fun(b=0,c=0)
   b. fun(a=1,b=0,c=0)
   c. fun(a=0)
   d. fun(1,c=2)

11. Find the output of the following code
```
def fun(x,y,z):
    x[y] = z
xy = {}
print(fun(xy,'1','z'))
```
   a. '1'
   b. 'z'
   c. None
   d. Error

12. For the following code, which one is correct?
```
from x import y
```
   a. x reside inside y
   b. y reside inside x
   c. None of the above
   d. Error

13. For the following code, which one is correct?
```
from x import y
```
   a. x is a module
   b. y is a module
   c. None of the above
   d. Error

14. Which of the following statement is in the correct syntax?
   a. from x import y
   b. from x import *
   c. import x
   d. All of the above

15. A function is defined inside
    a. Another Function
    b. Module
    c. Class
    d. All of the above

16. Find the output of the following code
    ```
    def pow(a,b=2):
        i = 1
        for j in range(b):
            i = i*a
        return i
    print(pow(5))
    print(pow(5,3))
    ```

    a. 5 and 125
    b. 25 and 125
    c. 5 and 25
    d. Error

17. Which of the following is a throw-away function?
    a. return
    b. parameter
    c. argument
    d. lambda

18. In function if yield is written in place of return, it is called
    a. function
    b. module
    c. argument
    d. generator

19. Which of the following is correct?
    a. Functions must have a return statement
    b. Generators have exactly one yield statement
    c. Functions have more than one return statement
    d. Generators have more than one yield statement

20. Which of the following is not correct?
    a. Lambda is a statement
    b. Lambda contains a return statement
    c. Lambda contains a block of statements
    d. All of the above

DESCRIPTIVE QUESTIONS

1. What are the advantages of a function?
2. What are the different ways of passing parameters to a function?
3. What is namespace?
4. What is meant by the scope of a variable?
5. What are the advantages of a module?
6. What is a lambda function? Explain with an example.
7. What is a generator? Explain the advantages with an example.
8. How do you pass variable length parameters?
9. What is recursion? What are the basic rules to implement a recursive function?
10. How to create a user-defined module?

PROGRAMMING QUESTIONS

1. Write a program that finds the GCD or HCF of two positive numbers.
2. Write a program that generates the factors of a positive Integer.
3. Write a program that generates the LCM of two positive integers.
4. Write a recursive function to implement binary search.
5. Write lambda functions that calculate the square of the range of numbers.

ANSWER TO MULTIPLE CHOICE QUESTIONS

1.	D	2.	D	3.	A	4.	B	5.	D
6.	B	7.	D	8.	D	9.	A	10.	B
11.	C	12.	B	13.	A	14.	D	15.	D
16.	B	17.	D	18.	D	19.	D	20.	D

9 File Handling

LEARNING OBJECTIVES

After studying this chapter, the reader will be able to:

- Open a file in a different mode and close the opened file
- Read a file and write into a file
- Understand file positioning using seek() and tell()
- Understand file operations using the "with" statement
- Understand serializing and de-serializing using pickling

9.1 INTRODUCTION

Until now we have learnt how to read data from the keyboard and how to display it on the screen. But every time this is not the situation. Sometimes we need to read from and write in the file also. In this chapter, we are going to discuss this.

9.2 FILE IO

Before we start reading from or writing into a file, we have to prepare the file and tell Python what we intend to do with the file.

9.2.1 FILE OPENING AND CLOSING

We have to tell Python what is the file format, if we intend to read from it, if we intend to write into it, or if we intend to append something at the end of it, etc. Before working on a file, we need to open it. For that we use the function *open()*. The general syntax is: ***File_Object = Open(File_Name[,mode] [,buffer])***.

Here *open* is the keyword and along with that there are two optional parameters:

- *File_name* is the name of the file we want to access.
- *mode* tells Python if we want to perform read(r) or write(w) or any other operation listed in table 9.1 with the file/
- *buffering* determines what kind of buffering will be used. If the value is set to 0, it indicates no buffering. If the value is set to 1, line buffering is performed. If the value is set to any positive number, then the buffering is done according to the size mentioned.

Assign the output of the open function to an object which can be any valid variable name. Every time we want to read or write from this file, we refer to that object

DOI: 10.1201/9781003219125-10

TABLE 9.1

File Access Mode

Mode	Description
r	Opens a file for reading only. This is the default mode.
rb	Opens a file for reading only in binary format
r+	Opens a file for both reading and writing.
rb+	Opens a file for both reading and writing in binary format.
w	Opens a file for writing only. Overwrites the file if the file exists. Creates a new file if it does not exist.
wb	Opens a file for writing only in binary format.
w+	Opens a file for both writing and reading. Overwrites the file if the file exists. Creates a new file if it does not exist.
wb+	Opens a file for both writing and reading in binary format.
a	Opens a file for appending. The file pointer is at the end of the file if the file exists. It does not read.
ab	Opens a file for appending in binary format.
a+	Opens a file for both appending and reading.
ab+	Opens a file for both appending and reading in binary format.

TABLE 9.2

File Property Functions

Methods	Descriptions
name()	Returns the name of the file we are dealing with now.
mode()	Returns the mode in which we have opened the file. It can be anything from table 9.1
closed	Returns a Boolean value to check whether the file is closed or not.
readable()	Returns a Boolean value to check whether the file is readable or not.
writable()	Returns a Boolean value to check whether the file is writable or not.

(variable name). There are different symbols that we can use for file access mode, each of them having a different purpose. The list is given in table 9.1.

It is highly recommended to close the file once the operations are over. The general syntax is: *file.close()*.

9.2.2 FILE PROPERTIES

It is always possible to find the properties of the file that we are dealing with. This can be done by a few file methods shown in table 9.2.

Example 9.1

```
f= open("MyTestFile.txt",'w')
print("File Name: ", f.name)
```

```
print("Final Mode: ", f.mode)
print("Is File Readable: ", f.readable())
print("Is File Writable: ", f.writable())
print("Is File Closed: ", f.closed)
x=f.close()
print("Is File Closed: ", f.closed)
```

Output

```
File Name: MyTestFile.txt
Final Mode: w
Is File Readable: False
Is File Writable: True
Is File Closed: False
Is File Closed: True
```

9.2.3 FILE READING

Once the file is open, we can read from a file by using the *read()* method or write into a file using the *write()* method. There are three different ways we can read a file:

- **file.read([size])**: It reads the entire file and returns a String. If the size is specified then read at most size bytes.
- **file.readline([size])**: It reads one line from the file and returns a String. If the size is specified then it reads that many number of bytes from the file.
- **file.readlines([size])**: It reads all the lines from the file and returns a list of lines. If the size is specified then it reads that many number of bytes from the file.

Example 9.2

```
fp= open("MyTestFile.txt", "r")
x= fp.read()
print(x)
fp.close()
```

Observe the 1st line, *fp* is a variable name which holds the output of an open function to this object. *MyTestFile.txt* is an existing text file. This file is opened in read mode which is indicated by *r*. In the 2nd line, if the file is opened successfully and the data in it is read successfully, all the data is assigned to a variable *x*. In the 3rd line, we have printed *x* and in the 4th line, we closed the file.

Output

```
I am Learning Python.
I am really Enjoying Learning Python
```

Example 9.3 File Operation (readline())

```
fp= open("MyTestFile.txt", 'r')
x= fp.readline()
print(x)
```

```
x= fp.readline()
print(x)
fp.close()
```

The *readline()* statement reads one line from the file and returns a String. We may have many lines in the text file. In order to read all the lines from the file, either we need to give many print statements or we need to use a loop and read lines until the end of the file.

Output

```
I am Learning Python.
I am really Enjoying Learning Python
```

Example 9.4 File operation(readlines())

```
fp= open("MyTestFile.txt", 'r')
x= fp.readlines()
print(x)
x= fp.readline()
print(x)
fp.close()
```

The *readlines()* method returns a list including the carriage return and each element of the list would be one line of the file. Observe that the last line does not contain a carriage return.

Output

```
['I am Learning Python.\n', 'I am really Enjoying Learning
Python']
```

9.2.4 FILE WRITING

Now let us discuss how to write into a file. In order to write into a file, we need to open the file in *w* mode. The filename that we have used for the *open* statement, if the file already exists, then we will overwrite it and if it does not exist, then Python creates a new file. There are three different ways we can write into a file.

- **file.write()**: It takes one or more lines of text data or a block of bytes and writes the data to the file.
- **file.writeline()**: It takes a list of strings and writes them into a file.
- **file.writerow()**: It takes a list and writes them into a file. This method is applicable to .csv files.

Example 9.5 File operation(write())

```
fp= open("MyTextFile.txt","w")
my_Str= input("Write Something!!!\n")
fp.write(my_Str)
fp.close
```

```
fp= open("MyTextFile.txt", "r")
print("Reading from File")
x= fp.readline()
print(x)
fp.close()
```

Observe the 1st line, we have used the file *MyTestFileW.txt* which is opened with write mode indicated by *w*. The *write()* method enables us to write into the file. If we run this program it will prompt us to enter something. We can see whatever is written into the file *MyTestFileW.txt* by opening it in *r* file access mode, as we described in an example earlier.

Output

```
Write Something!!!
File Handling is easy in Python
Reading from File
File Handling is easy in Python
```

9.2.5 BINARY FILE READING/WRITING

Like any text file, it is common to read or write binary files like images, video files, audio files, etc. From table 9.1, we can see that the binary file reading and writing is done with mode *rb* and *wb* respectively.

Example 9.6 Binary file handling

```
f1= open("pic.jpg","rb")
f2= open("newpic.jpg","wb")
bytes= f1.read()
f2.write(bytes)
print("New Image is available with name: newpic.jpeg")
f1.close()
f2.close()
```

Observe the first line, *f1=open("Pic.jpg","rb")*, this indicates the file *pic.jpg* is opened with mode *rb*, (reading a binary file). The second line *f2=open("newpic.jpg","wb")*, indicates *newpic.jpg* is opened with mode *wb* (writing binary file). The third line indicates we are reading the first file *pic.jpg*, and the 4th line indicates writing the data that we read now in *newpic.jpg*. Here the *newpic.jpg* is created at the same location as *pic.jpg*.

Output

```
New Image is available with name: newpic.jpg
```

9.2.6 CSV FILE READING/WRITING

CSV refers to Comma Separated Value. In this file format the fields of a record are separated by a comma. In other words, we use a comma as a delimiter between fields of a record. In order to handle a CSV file, we need to import *csv*.

Example 9.21 CSV file handling

```
import csv
with open("student.csv", "w", newline='') as f:
  w= csv.writer(f)
  w.writerow(["S.No", "Roll No", "Name"])
  n= int(input("Enter number of students:"))
  for i in range(n):
    sno= input("Enter student no.:")
    sroll= input("Enter student roll no.: ")
    sname= input("Enter student name: ")
    w.writerow([sno,sroll,sname])
print("Total Students data written to csv file successfully.")
```

Output

```
Enter number of students:2
Enter student no.:1
Enter student roll no.: 501
Enter student name: Sushobhan Raj
Enter student no.:2
Enter student roll no.: 502
Enter student name: Shreya Shree
Total Students data written to csv file successfully.
```

9.3   FILE POSITION

There are two methods that can be used to position the file pointer to any position in the file and also ask Python, where is the current position. These are shown in table 9.3.

TABLE 9.3
File Position Methods

Methods and Functions	Description
file.seek()	Changes the file object's position.
file.tell()	Returns an Integer giving the file object's current position in the file.

Example 9.7 File position(seek())

```
fp= open("MyTestFile.txt", "r")
fp.seek(10)
x= fp.read()
print(x)
fp.close()
```

Observe that in the 2nd line we have written *fp.seek(10)*. It instructs the Python interpreter to start reading the file from position 10 rather than starting from position 0. The file starts reading from position 11. We can see that the first ten characters which were there are not printed.

Output

```
ing Python.
I am really Enjoying Learning Python
The tell() method is used for giving the current location in
the file.
```

Example 9.8 File position(tell())

```
fp= open("MyTestFile.txt", "r+")
pos= fp.tell()
print("The Position at the beginning is ", pos)
fp.seek(10)
pos= fp.tell()
print("The position at the end is ", pos)
fp.close()
```

Observe the second line of the code *pos = fp.tell()*. After opening the file in read mode, we have taken the current position. As evident from the output, it is 0. In the 4th line we have *fp.seek(10)*. It means before starting reading the file we have skipped 10 positions. Now the file starts reading from 11th position till the end of the line. Again, in line number five we have taken the position of the file pointer by *pos = fp.tell()*. Now the print statement gives 10 as the current location.

Output

```
The Position at the beginning is 0
The position at the end is 10
The same operations can also be done while writing into a
file.
```

Example 9.9 File position(seek())

```
fp= open("MyTestFile.txt", "r+")
fp.seek(3)
fp.write("...XXXXXX...")
fp.close()
```

Observe that in the 2nd line of the code we have *fp.seek(3)*. It means the current file pointer points to position 4. From the 4th location "...*XXXXXX*..." is written onto the file. From the 4th location onwards the String is overwritten until the end. The rest of the file remains as it is. It means with mode *r+* we overwrite and do not insert into the file.

Output

```
I a...XXXXXX...ython.
I am really Enjoying Learning Python
```

There are several file methods available. We have already discussed some of them and some more are listed here.

9.4   SOME MORE FILE OPERATIONS

There are many more operation utilities available which are useful for file manipulation. We discuss only a few of them in table 9.4. For more functions, please visit www.python.org/doc/.

TABLE 9.4
Few More File Operation

Methods	Description
file.close()	Closes a file which is open. Except for opening a file, no other operations can be performed if the file is closed.
file.flush()	It flushes the internal memory.
file.fileno()	It returns the file descriptor of the stream as an Integer which is the file descriptor to request I/O operations from the operating system.
next(file)	It returns the next line.
f.truncate(n)	It truncates the file size at most up to 'n' bytes.

Example 9.10 File operations-fileno(), next(), isatty()

```
f= open("MyStudentList.txt", 'r')
print("The file No is ", f.fileno())
print("The First line is: ", next(f))
print("The next line is: ", next(f))
print("The file is connected to terminal: ", f.isatty())
f.close()
```

Output

```
The file No is 3
The First line is: 526      KARTHIK
The next line is: 527       SUCHITHA
The file is connected to terminal: False
```

Example 9.11 File operation truncate

```
import os
size= os.path.getsize("MyTestFile.txt")
print("The file size originally: ", size, "bytes")
f= open("MyTestFile.txt", "w")
```

```
print("The file is truncated: ")
f.truncate(100)
f.close()
size= os.path.getsize("MyTestFile.txt")
print("The file size after truncation: ",size,"bytes.")
```

The truncate method is used while the file is opened in write mode. The method *os.path.getsize()* returns the file size. The method *f.truncate(100)* makes the size of the opened file maximum up to 100 bytes.

Output

```
The file size originally: 55 bytes
The file is truncated:
The file size after truncation: 100 bytes.
```

Now observe that when we run the program for the next time the original file size is 100 bytes. By changing 6th line of the above program to *f.truncate(10)*, the size of the file is changed to 10 bytes.

Output

```
The file size originally: 100 bytes
The file is truncated:
The file size after truncation: 10 bytes.
```

9.5   OPERATING SYSTEM–RELATED OPERATIONS

There are some operating system–related methods that can also be used for efficient file manipulation. A few of the operating system–related operations which are frequently used are listed in table 9.5. For a full list of operations related to the operating system, please visit docs.python.org/3/library/os.html.

TABLE 9.5
Methods Related to Operating Systems

Methods	Description
os.chdir()	Change Directory.
os.getcwd()	Returns the current working directory.
os.rename()	Renames a file.
os.remove()	Deletes a file.
os.mkdir()	Make a new directory.
os.rmdir()	Deletes the entire directory.
os.is_file()	Returns True if it is filename otherwise returns False.
os.is_dir()	Returns True if it is directory name otherwise returns False.

Example 9.12

```
import os
print(os.getcwd())
os.chdir('C:\IoT')
print(os.getcwd())
```

It can be seen from the output that the current working directory changed after we call *chdir()* method. The *chdir()* method takes a parameter in the form of a String, and it simply changes the directory returning *None*. The method *getcwd()* returns the directory we are now in.

Output

```
C:\Users\RakeshN\AppData\Local\Programs\Python\Python37-32\f
ilehandling
C:\IoT
```

Example 9.13

```
import os
file_name= input("Enter the file name to be renamed: ")
file_name_renamed= input("Enter the new file name:")
os.rename(file_name, file_name_renamed)
print("Renaming Successful")
```

The *os.rename()* method renames a filename without returning anything. Both the old and new paths and filenames are needed as parameters in order to rename. If we don't supply the path name it will try to locate the file in the current working directory. If the file is not found, it will raise *FileNotFoundError*.

Output

```
Enter the file name to be renamed: pic.jpg
Enter the new file name:pic1.jpg
Renaming Successful
```

Example 9.14

```
import os
os.chdir('c:\TC')
dir_name= input("Enter the Directory Name to Create:")
os.mkdir(dir_name)
print(os.listdir('c:\TC'))
dir_name= input("Enter the Directory Name to Remove: ")
os.rmdir(dir_name)
print(os.listdir('c:\TC'))
```

Observe that by using *os.mkdir()* we have created a folder named *NewFolder* in the current working directory. It takes a directory name with a path name as a parameter and returns *None*. Use the *os.rmdir()* method to remove the directory name. It takes a directory name with a path name as a parameter and returns *None*.

Output

```
Enter the Directory Name to Create:abc
['abc', 'BGI', 'BIN', 'CLASSLIB', 'DOC', 'EXAMPLES',
'FILELIST.DOC', 'INCLUDE', 'LIB', 'Like on Facebook.url',
'README', 'README.COM', 'turboc7.blogspot.com.url',
'unins000.dat', 'unins000.exe']
Enter the Directory Name to Remove: abc
['BGI', 'BIN', 'CLASSLIB', 'DOC', 'EXAMPLES', 'FILELIST.
DOC', 'INCLUDE', 'LIB', 'Like on Facebook.url', 'README',
'README.COM', 'turboc7.blogspot.com.url', 'unins000.dat',
'unins000.exe']
```

Example 9.15

```
import os
print(os.remove('pic1.jpg'))
print(os.remove('pic1.jpg'))
```

This method deletes the file and returns *None*. If the path is not mentioned it deletes the file in the current working directory. If the file is not available it raises *FileNotFoundError*.

Output

```
None
Traceback (most recent call last):
  File "C:\Users\RakeshN\AppData\Local\Programs\Python\Py
  thon37-32\filehandling\chap-9.py", line 58, in <module>
    print(os.remove('pic1.jpg'))
FileNotFoundError: [WinError 2] The system cannot find the
file specified: 'pic1.jpg'
```

Example 9.16

```
import os
filename= input("Enter file name: ")
if os.path.isfile(filename):
  print("filer found")
else:
  print("file Not Found")
dirname= input("Enter directory name: ")
```

```
if os.path.isdir(dirname):
  print("file Found")
else:
  print("file not found")
```

Output

```
Enter file name: pic1.jpg
filer found
Enter directory name: abc
file not found
```

9.6   THE "WITH" STATEMENT

We must close the file handler every time we open it. Often it is difficult to remember closing it once the work is done with the file. In order to avoid this, Python provides a solution. The "with" statement automatically closes the connection once the work is over.

The general syntax of opening a file is: ***with open(file_name) as filepointer_name***.

Example 9.17

```
with open("MyTextFile.txt", "r") as fp:
  x= fp.read()
  print(x)
```

In this example, we can see that it is starting with the keyword *with* followed by *open()*. For reading a file, the file opening mode *r* is optional. While writing into a file, the file opening mode *w* must be specified.

Output

```
File Handling is easy in Python
```

The *with* statement supports multiple file handlers to work with at the same file.

Example 9.18

```
with open("MyTestFile.txt") as fp1, open("MyTestFile.txt")
as fp2:
  # Action to perform
  # with the opened files
```

9.7   PICKLING

Pickling refers to the serializing and de-serializing of Python objects. The process of converting a Python object to a byte stream is called serializing and the reverse process is called de-serializing. In order to use pickling we need to import pickle modules. Python objects like strings, lists, tuples and dictionaries can be used for pickling. There are two methods (*dump()* and *load()*) that come with pickle modules.

Example 9.19 Pickling

```
import pickle
with open("MyTextFile.txt") as fr, open("MyTestFile.
txt", "wb") as fw:
  x= fr.read()
  pickle.dump(x,fw)
  print("Converted to byte stream successfully.")
```

The 1st line imports the *pickle* module. The *pickle.dump()* method takes two parameters. The first parameter is the Python object to be dumped and the second parameter is the file pointer of the file to which the Python object is to be written. The file to which the Python object is to be dumped must open in *wb* mode. After executing this code, a file *MyTestFile1.txt* will be created in the current directory.

Output

```
Converted to byte stream successfully.
```

Example 9.19 Pickling

```
import pickle
with open("MyTestFile.txt", "rb") as fr:
  x= pickle.load(fr)
  print(x)
```

In order to retrieve the data from the dumped file, *pickle.load()* method is to be used. It will take only one parameter. This is a file pointer which is opened in *rb* mode. We can assign a variable to it which can be printed later on.

Output

```
File Handling is easy in Python
```

WORKED OUT EXAMPLES

Example 9.22 Write a program in Python that will read only the last three records from a student file which is a .txt file

```
from itertools import islice
with open("MyStudentList.txt") as f:
  line = f.readline()
  i= 0
```

```
while line:
  line = f.readline()
  i += 1

f.seek(i-2)
for line in islice(f, i-2, None):
  print(line)
```

Example 9.23 Write a program in Python that will read only the first two records from a student file which is a .txt file

```
from itertools import islice
with open("MyStudentList.txt") as f:
  for line in islice(f,2):
    print(line)
```

Example 9.24 Write a program in Python that will find the longest name from a student file which is a .txt file

```
with open("MyStudentList.txt") as infile:
  words = infile.read().split()
  mx = 0
  for i in words:
    if len(i) >= mx:
      mx= len(i)

  for j in words:
    if len(j) == mx:
      print(j)
```

MULTIPLE CHOICE QUESTIONS

1. The two different file open modes are
    a. Text and image
    b. Binary and text
    c. Binary and ternary
    d. Image and binary

2. The method able to read data from a file into a byte array object is
    a. readin()
    b. readout()
    c. readinto()
    d. readloud()

3. The readline() method returns
    a. A list
    b. A tuple
    c. A String
    d. A dictionary

4. How does readline() react to end-of-file?
   a. It raises an exception
   b. It returns −1
   c. It raises an error
   d. It returns an empty String

5. Look at the following code
   X=s.read(1)
   If s is a stream opened in read mode, X will

   a. Read 1 character from the stream
   b. Read 1 character from the buffer
   c. Read 1 line from the stream
   d. Read 1 word from the buffer

6. To append to a text file which mode is used?
   a. w
   b. ab+
   c. a
   d. ab

7. When a file is opened in read mode
   a. it must exist (exception will be raised otherwise)
   b. it need not exist
   c. it will be deleted if it exists
   d. it creates if it does not exist

8. Which of the following is correct for creating a text file? If the file does not exist, the command should create and if the file exists then it should overwrite the text file.
   a. open('abc.txt','r')
   b. open('abc.txt','r+')
   c. open('abc.txt','w+')
   d. open('abc.txt','w')

9. Consider the code
   f = open('abc.txt')
   f.readall()
   Which exception will be raised if the file does not exist?

   a. EOFError
   b. SystemError
   c. SyntaxError
   d. FileNotFoundError

10. Which of the following statement is true?
    a. When you open a file for reading, if the file does not exist, an error occurs
    b. When you open a file for writing, if the file exists, the existing file is overwritten with the new file

  c. When you open a file for writing, if the file does not exist, a new file is created

  d. All of the mentioned

11. To read the next line of the file from a file object infile, we use
  a. infile.read(2)
  b. infile.read()
  c. infile.readline()
  d. infile.readlines()

12. tell() method is used in Python to
  a. Tell you the current position within the file
  b. Tell you the file is opened or not
  c. Tell you the end position within the file
  d. None of the above

13. What is pickling?
  a. It is used for object serialization
  b. It is used for object deserialization
  c. None of the above
  d. All of the above

14. Find the correct syntax for file.writelines()
  a. fileObject.writelines()
  b. file.writelines(sequence)
  c. fileObject.writelines(sequence)
  d. None of the above

15. To close an opened file, which command is issued in Python?
  a. close()
  b. stop()
  c. end()
  d. closefile()

16. What is the difference between the r+ and w+ modes?
  a. In r+ the pointer is initially placed at the beginning of the file and the pointer is at the end for w+
  b. In w+ the pointer is initially placed at the beginning of the file and the pointer is at the end for r+
  c. Depends on the operating system
  d. No difference

17. To rename a file, which of the following command is used?
  a. fp.name = 'new_name.txt'
  b. os.rename(existing_name, new_name)
  c. os.rename(fp, new_name)
  d. os.set_name(existing_name, new_name)

18. How do you change the file position from the start?
    a. fp.seek(offset, 0)
    b. fp.seek(offset, 1)
    c. fp.seek(offset, 2)
    d. None of the mentioned

19. Which of the following is not an attribute related to a file object?
    a. closed
    b. mode
    c. name
    d. rename

20. The seek() method
    a. Tells the current position within the file
    b. Indicates that the next read or write occurs from that position in a file
    c. Determines if you can move the file position or not
    d. Moves the current file position to a different location at a defined offset

DESCRIPTIVE QUESTIONS

1. What are the differences between append and write methods?
2. Why it is better to use "with" while dealing with files?
3. What is pickling?
4. What are the different file opening modes?
5. Explain the seek() and tell() methods.

PROGRAMMING QUESTIONS

1. Write a program in Python that will count the number of records from a student file which is a .txt file.
2. Write a program in Python that will read only the first *n* number of records from a student file which is a .txt file without using any other modules.
3. Write a program in Python that will read only the last *n* number of records from a student file which is a .txt file without using any other modules.
4. Write a program in Python that will print a random line from a student file which is a .txt file.

ANSWER TO MULTIPLE CHOICE QUESTIONS

1.	B	2.	C	3.	A	4.	D	5.	A
6.	C	7.	A	8.	C	9.	D	10.	D
11.	C	12.	A	13.	A	14.	C	15.	A
16.	A	17.	B	18.	A	19.	D	20.	D

Section II

Object Oriented Concepts in Python

10 Classes and Objects

LEARNING OBJECTIVES

After studying this chapter, the reader will be able to:

- Understand object-oriented concepts
- Define and use classes and objects
- Define and use different types of variables
- Define and use different types of methods
- Define and apply inner class and know how to access attributes of an inner class
- Use some special methods and attributes associated with a class

10.1 INTRODUCTION TO OBJECT-ORIENTED PROGRAMMING

Python is a universal tool for both object and procedural programming. It can be successfully utilized in both spheres. The procedural style of programming was the dominant approach to software development for decades where functions can use data, but not vice versa. The object-oriented approach is useful when applied to big and complex projects carried out by large teams consisting of many developers. In programming, we try to solve real-world problems with the help of a virtual world solution. In the real world, everything is an object.

Example 10.1

Let us discuss an employee as an object. If we are talking about a company, we want some work to be done and to do the work we need employees. Every employee is an object. Employee object may need some other object to do work with it. For example, a computer, a pen, some tools, etc. Each tool is an object. Each object comprises attributes and behaviors. Attributes are what it has (the properties) that describe the object, and behavior is what it does (the functions). In other words, objects are those which can be described and which can do something.

The object is something that can store some data and have some behavior. We store data in a variable and if we want to define the behavior, we need to use methods. Functions in object-oriented programming are called methods. Generally, a function that is invoked with the help of a dot (.) operator is known as a method.

An employee has attributes such as name, age, employee_id, date_of_birth, date_of_joining, etc. The behavior is the task performed by the employee. Behavior is defined by action. Figure 10.1 depicts this scenario.

Let us take another real-world object, a fan. Assume a factory that produces fans. Every fan of the same model has the same look and it works in the same way.

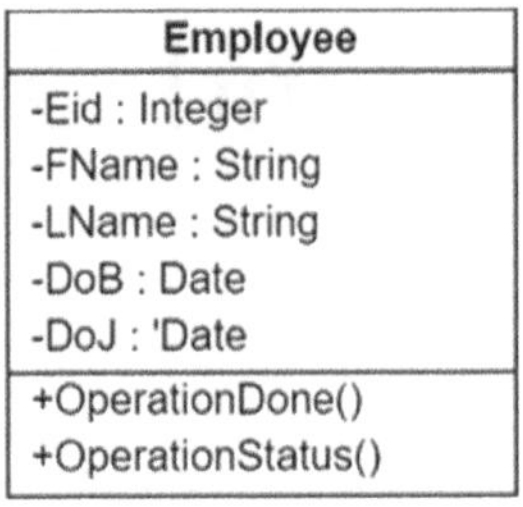

Employee
-Eid : Integer
-FName : String
-LName : String
-DoB : Date
-DoJ : 'Date
+OperationDone() +OperationStatus()

FIGURE 10.1   Object attributes

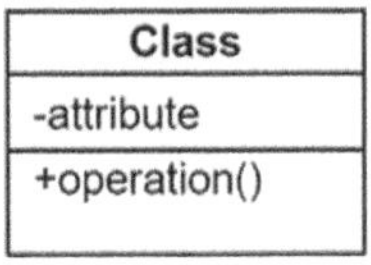

Class
-attribute
+operation()

FIGURE 10.2   Creation of a class

It is because of the design. Based on a particular design the factory produces fans. The design of an object in object-oriented programming language is known as class. It means the class is designed for objects. When we discuss an employee as an object, all employee(s) have the same attribute; the values in these attributes may vary. All the employee(s) perform certain work. It means to create an employee object we need a template, which is a *class*. This is shown in figure 10.2.

One more term we use for the *object*, is an *instance*. An object is an instance of a class. So, if we have a class, we can create as many objects as we want. In fact in Python, everything is a class.

Example 10.2

```
a= 5
print(type(a))
a= 5.5
print(type(a))
a= "Python"
print(type(a))
```

Output

```
<class 'int'>
<class 'float'>
<class 'str'>
```

Observe here, the output of first print statement is *<class 'int'>*. Similarly, observe lines with output *<class 'float'>* and *<class 'str'>*. It means even the inbuilt datatypes are also a class in Python.

10.2   CLASS AND OBJECT

The classes defined at the beginning are too general. The class we define has nothing to do with the objects.

10.2.1   DEFINING A CLASS

We know that a class is a template of an object. Unless the class is defined properly we cannot get a proper object.

The general syntax for a class is:

class ClassName:
    class documentation String'
    class_suite

In a class notice that:

- The definition begins with the keyword *class.*
- The keyword *class* is followed by an *identifier* that is the name of the class followed by a colon (:).
- The *class_suite* is described inside the class.
- Everything under *class_suite* should be properly tabbed.
- The *class_suite* consists of two things: the *attributes* and the *behavior.*

Attributes are the variables and behaviors are the methods that are described in the class. Every class is divided into three parts:

1. A class has a *name* that uniquely identifies within its home namespace.
2. A class has a set of individual *properties.*
3. A class has a set of *operations.* It is the ability to perform specific activities.

It is possible to leave the *class_suit* empty, a class with no attributes and no methods.

Example 10.3 Class definition

```
class Employee:
  pass
```

In this example, the *class_suit* contains pass. The *pass* is a keyword that fills the class with nothing. It doesn't contain any attribute or any method, it is just a placeholder.

Example 10.4 Class definition

```
class Employee:
  WorkPlace= 'xxxyyyzzz'
```

```
def __init__(self):
  self.id= 502
  self.name= 'Rakesh'
def job(self):
  self.position= 'Professor'
```

In this example, the *class_suit* contains attributes (variables) and methods (behaviors). Here *WorkPlace, id* and *name* are attributes, whereas *__init__()* and *job()* are methods.

Example 10.5 Class definition

```
class Employee:
  def Emp_Details(self):
    print('Employee has id: ', 502, ', Name: ', 'Ayaansh Gupta')
```

In this example we have defined a class named *Employee*. In the *class_suit* a method named *Emp_Details()* is defined and no attributes are defined.

There are different types of attributes and methods in a class. We will discuss all of them later in this chapter.

10.2.2 GENERATING AN OBJECT

We mentioned that class definition is just a template or a design of all the objects. So, without creating any object, having a class is useless. In other words, an object is an instance of a class as shown in figure 10.3.

Every employee object has an Eid, FName, LName, DoB and DoJ which is different from other objects. It means any individual employee has unique properties, which makes the employee object unique.

Normally, to describe any object we use (although it may not always work):

- A *noun* that defines the object's name.
- An *adjective* that defines the object's property.
- A *verb* that defines the object's operation.

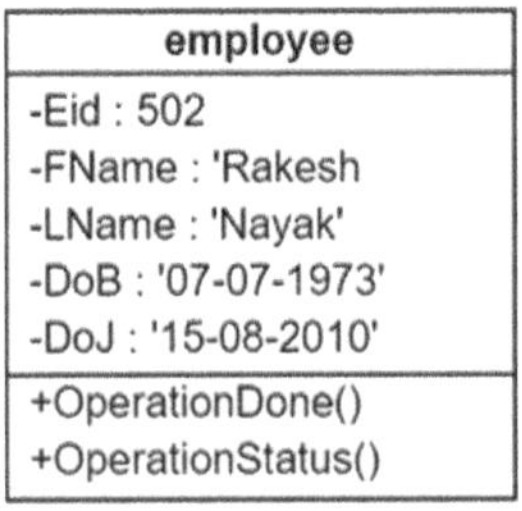

employee
-Eid : 502
-FName : 'Rakesh'
-LName : 'Nayak'
-DoB : '07-07-1973'
-DoJ : '15-08-2010'
+OperationDone()
+OperationStatus()

FIGURE 10.3   Generation of an object

The existence of a class does not mean that any of the compatible objects will auto-matically be created. The class itself can't create an object; we have to create it ourselves. The already defined class becomes a tool that can create new objects. The tool has to be used explicitly. We need to assign a variable (object name) to store the newly created object of that class. The general syntax for creating an object is:

Object_name = Class_name()

The returned object is an instance of the class we called. When we assign a class using the functional notation, the newly created object is equipped with everything that the class possesses. The act of creating an object of the selected class is also called *instantiation* (as the object becomes an instance of the class).

Example 10.6 Object Creation and invocation of methods

```
Emp = Employee()
Emp.Emp_Details()
Employee.Emp_Details(Emp)
```

This example is an extension of example 10.5. The 1st line creates an object *Emp* from the defined class *Employee*. There are two different ways we can use the methods defined in a class.

METHOD 1

Observe the 2nd line, *Emp.Emp_Details()*. We are using the object itself to call the method. Remember that *Emp* is an object of the class *Employee*. *Emp_Details()* method of object *Emp* takes the parameter *self*. It means we want to access the method which is a part of object *Emp*.

METHOD 2

We know that *Emp_Details()* is a method defined in the class Employee. When we want to access the methods defined inside the class, we must pass the object name as a parameter. There may be many objects that are created from the same class. To indicate which object is accessing the method, we need to pass the object name as an argument. In the 3rd line, *Employee.Emp_Details(Emp)*, the method *Emp_Details()* is accessed by object *Emp*, which is passed as an argument to access the method defined in the class *Employee*. In either method, the output is the same.

Output

```
Employee has id: 502 , Name: Ayaansh Gupta
Employee has id: 502 , Name: Ayaansh Gupta
```

10.2.3   THE _ _ INIT _ _ () , _ _ NEW _ _ () AND _ _ DEL _ _ () METHODS

Two methods, __init__() and __new__() are crucial for initializing the values and serving as constructors for objects respectively, while __del__() serves as a destructor.

10.2.3.1   __init__()

The main objective of this method is initialization. The general syntax is: __*init*__ *(self [,arguments])*.

Notice that:

- The method name is *init* and it is prefixed and suffixed with two underscores (__).
- The parameter is *self* in the method.
- Along with *self* a list of arguments may also be passed.
- It does not return anything.

While calling the method, we do not pass the parameter *self*. Python provides it automatically. The *self* argument refers to the object itself. That is, the object that has called the method. This means, __*init*__() method takes no arguments and it should accept *self*. Similarly, a method defined to accept one argument will actually take two, that is self, and the argument.

This is the first method that gets executed automatically when an object is created. Every time an object is created, the __*init*__() method is called. This method is useful to initialize variables of the class object.

An explicit return statement should not be written for the *init()* method because the instance object is automatically returned after the __*init*__*(self)* is called. Otherwise, there is a conflict of interest because only the instance should be returned. Attempting to return any object other than *None* will result in a *TypeError* error.

There are two different ways of assigning values to the variable in an object.

Method 1

Inside the __*init*__() method, we can assign the value to the variable statically. Assume that the starting salary is fixed for all employees who join the organization. When the employee object is created at that time itself we can assign a value for salary.

Example 10.7

```
class Employee:
  def __init__(self):
    self.salary= 50000
Emp1= Employee()
print("Employee Salary", Emp1.salary)
```

Observe the method __*init*__*(self)*. Here a value 50000 is assigned to *self.salary*. When we create the object *Emp1*, we have not explicitly called the __*init*__() method. A variable named *salary* is created automatically and 50000 is assigned to it. The *self* refers to the object calling the method that is *Emp1* in this case.

Method 2

A value can be passed to the __init__() method that can assign the value to the variable dynamically.

Example 10.8

```
class Employee:
  def __init__(self,sal):
    self.salary = sal
Emp1 = Employee(50000)
print("Employee Salary", Emp1.salary)
```

Observe the __init__(self, sal) method. Apart from the parameter *self*, one more parameter *sal* is also present. While creating the object, we passed *50000* as an argument in *Emp1 = Employee(50000)*. This value is assigned to *self.salary*. This is shown in figure 10.4.

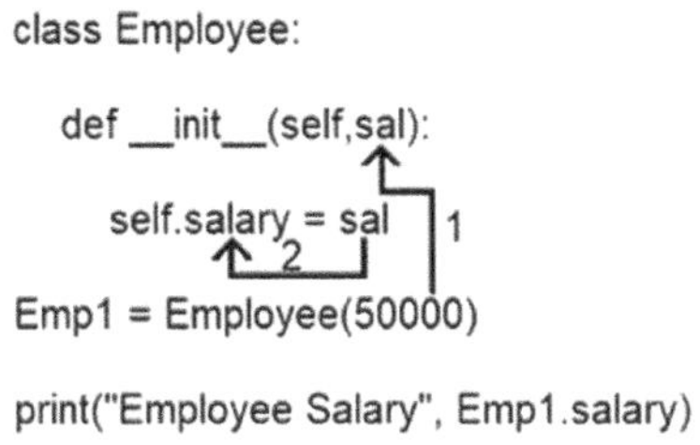

class Employee:

def __init__(self,sal):

self.salary = sal   1

Emp1 = Employee(50000)

print("Employee Salary", Emp1.salary)

FIGURE 10.4 Initialization of variables

In either case, the output remains the same. In the first method, for the entire object, the value of salary is by default 50000 whereas the second method gives some flexibility in assigning value by passing it as an argument.

Output

```
Employee Salary 50000
```

Example 10.9

```
class Employee:
  def __init__(self,sal):
    self.salary= sal
Emp1= Employee(50000)
Emp2= Employee(50500)
print("Employee 1 salary ", Emp1.salary)
print("Employee 2 salary ", Emp2.salary)
```

Observe that we have created two objects i.e., *Emp1* and *Emp2*. While creating the objects we passed values 50000 and 50500 respectively. We can see the output, for each object, the __init__() method is called only once with different values. So is the output.

Output

```
Employee 1 salary 50000
Employee 2 salary 50500
```

The name *self* is not a keyword, it is just a reference to the current object. So in place of *self* we can use any other variable. But it is a convention to use *self* to refer to the current object. Throughout this book, we will follow the convention of using *self*.

Example 10.10

```
class Employee:
  def __init__(a):
    a.salary= 50000
Emp1 = Employee()
print("Employee Salary", Emp1.salary)
```

Observe that in place of *self* we have used a valid variable name *a* and still the program worked. Which shows we can use any other variable name for *self*.

Output

```
Employee Salary 50000
```

10.2.3.2   __new__()

The main purpose of this method is object creation. The general syntax is:

object.__new__(cls[, *args, **kwargs])
  or
super(class_name,cls).__new__(cls [, *args, **kwargs])

Notice that:

- The method name is *new* and it is prefixed and suffixed with two under-scores (__).
- The parameter is *cls* in the method.
- Along with *cls* a list of arguments may also be passed.
- It returns an object.

It is a static method that takes *cls* as the first parameter and it returns a new object instance. This method can be thought of as a *constructor*. This method is called when the new class is ready to instantiate itself. This method is used to customize the creation of objects.

Example 10.11

```
class Employee(object):
  def __new__(cls, id, name):
    if 500<id<600:
      return object.__new__(cls)
    else:
      return None
  def __init__(self, id, name):
    self.id = id
    self.name = name
  def __str__(self):
    return '{0}{1}'.format(self.__class__.__name__,
    self.__dict__)
Emp1= Employee(501, 'Aruna Kranthi')
print(Emp1)
Emp2= Employee(224, 'Ramesh')
print(Emp2)
```

Observe that in the __new__() method the parameters are *cls, id* and *name*. The parameter *cls* indicates that __new__() is a static method. The object is only created when the value of *id* is between 501 and 599 (both inclusive). When the object is created, it returned the object along with its data, and it returned *None* when the object is not created.

Output

```
Employee{'id': 501, 'name': 'Aruna Kranthi'}
None
```

The __new__() method is always called before the __init__() method. It is because the initialization of value is possible only when there is an instantiation. The __init__() is called automatically every time an instance of a class is returned by __new__(). This instance is passed to the __init__() method passing *self* as a parameter.

It is not required to have the __init__() method in order to initialize some value to a variable when we are using the__new__() method.

Example 10.12

```
class Employee(object):
  def __new__(cls, id, name):
    if 500< id< 600:
      self= object.__new__(cls)
      self.id = id
      self.name = name
      return self
    else:
      return None
```

```
     def __str__(self):
       return '{0}{1}'.format(self.__class__.__name__,
       self.__dict__)
Emp1 = Employee(501, 'Ayaansh')
print(Emp1.id, Emp1.name)
Emp2 = Employee(224, 'Ramesh')
print(Emp2)
```

Observe that we have created objects based on the value of *id* in the *__new__()* method. When the object creation is successful the value of *id* is assigned to *self.id* and the value of *name* is assigned to *self.name* and then *self* is returned. An object *Emp1* is created and the attributes can be printed by *print(Emp1.id,Emp1.name)*. As *Emp2* is not created it will return *None* and if we want to print the value by *Emp2.id* and *Emp2.name* it will raise an *AttributeError*. So, it is always advisable to use overloading *__str__()* method when we are creating objects using *__new__()* method.

Output

```
501 Ayaansh
None
```

10.2.3.3   __del__()

The purpose of a destructor is to perform a cleanup activity before the garbage collector destroys the object. The general syntax is:

object.__del__(self):
 # body of the destructor

This method is called when all references to an object are deleted. The concept of removing unreferenced objects from the memory is called garbage collection. When the reference count to an object is zero, Python interpreter calls the garbage collector to free that object from the memory.

Example 10.13

```
class Test:
  def __init__(a):
    print("Object is Created")

  def __del__(self):
    print("Object is Deleted")

obj1 = Test()
print("Program End")
```

Observe the simple program. The object *obj1* creates an instance of the class *Test*. In other words, the reference to the object created by class *Test* is *obj1*. As long as the reference exists the object is not destroyed.

Output

```
Object is Created
Program End
```

To destroy an object either we need to call the *del* method explicitly or we need to delete the reference.

Example 10.14

```
class Test:
  def __init__(self,a):
    self.no = a
    print("Object ",a, " is Created")

  def __del__(self):
    print("Object is Deleted")

obj1 = Test(1)
obj1 = 5

obj2 = Test(2)
del obj2

print("Program End")
```

Observe that the object created by *obj1 = Test(1)* is destroyed. Here the destructor method is called automatically as soon as *obj1* references to 5. Similarly, the object created by *obj2 = Test(2)* is destroyed when *del obj2* is executed.

Output

```
Object 1 is Created
Object is Deleted
Object 2 is Created
Object is Deleted
Program End
```

If more than one reference is pointing to the same object, then the destructor cannot be called unless all the references are removed.

Example 10.15

```
class Test:
  def __init__(self,a):
    self.no = a
    print("Object ",a, " is Created")

  def __del__(self):
    print("Object is Deleted")
```

```
obj1 = Test(1)
obj2 = obj1
del obj1
print("Program End")
```

Observe that obj1 and obj2 are pointing to the same object created by *obj1* = *Test(1)*. The statement *del obj1* has not called the destructor as the object is still existing which is referenced by *obj2*.

Output

```
Object 1 is Created
Program End
```

Example 10.16

```
class Test:
  def __init__(self,a):
    self.no = a
    print("Object ",a, " is Created")

def __del__(self):
  print("Object is Deleted")

obj1 = Test(1)
obj2 = obj1

del obj1
del obj2
print("Program End")
```

Here, observe that we have two statements *del obj1* and *del obj2*. Both the references to the same object are deleted and as a result the destructor is called.

Output

```
Object 1 is Created
Object is Deleted
Program End
```

10.3   VARIABLES AND METHODS

Variables and methods are an important part of any class. In this section we are going to discuss different types of variables and methods and the way we access them.

10.3.1   VARIABLES (ATTRIBUTES)

There are two types of variables – *object variables* and *class variables*. The object variables are owned by each object whereas the class variables are owned by the class.

10.3.1.1 Object Variables (Instance Variables)

Instance variables are the variable that change from object to object. Every object creates a separate copy of the variable. We can create an instance variable in three different ways.

- Creating an instance variable inside __init__() using self variable.
- Creating an instance variable outside the class by using the object reference variable.
- Creating an instance variable inside the instance method by using the self variable.

10.3.1.1.1 Creating an Instance Variable Inside a __init__() by Using the Self Variable

In this method, we declare instance variables inside a __init__() by using *self*. Once we create an object, automatically these variables will be added to the object.

Example 10.13

```
class Employee:
  def __init__(self):
    self.id= 265
    self.name= 'Sushobhan Raj'
Emp1= Employee()
Emp2= Employee()
print("Employee has id: ", Emp1.id, ", name: ", Emp1.name)
print("Employee has id: ", Emp2.id, ", name: ", Emp2.name)
```

Observe here that *self.id* and *self.name* are instance variables. By default the values are *265* and *'Sushobhan Raj'* respectively for all the objects. As many objects we create, for all the objects we get the same values for id and name.

Output

```
Employee has id: 265 , name: Sushobhan Raj
Employee has id: 265 , name: Sushobhan Raj
```

10.3.1.1.2 Creating an Instance Variable Outside the Class by Using the Object Reference Variable

The values of the variables (attributes) can also be changed after the object is created. It is done by referencing the object. The general syntax is:

Object_name.variable_name = value.

Example 10.14

```
class Employee:
  def __init__(self):
```

```
   self.id= 502
   self.name= 'Rakesh'
Emp1= Employee()
Emp2= Employee()
Emp1.id= 501
Emp1.name= "Ayaansh"
Emp1.phone= 89746827937
print("Employee has id", Emp1.id,", name ",Emp1.name,"
having phone number ", Emp1.phone)
print("Employee has id", Emp2.id,", name ",Emp2.name)
```

Observe that the values *502* and *Rakesh* are assigned to the variable id and name respectively for all objects created. The values of these variables can be changed or a new variable can also be added after the object is created. We can do so by mentioning for which object the value of which variable changed or which new variable is added. Assign a new value to it. Here we can observe that only the values of variables in *Emp1* are changed to *501* and *Ayaansh*. Along with that, one more variable *phone* is also added to the object. If we try to access the variable *phone* for *Emp2*, it will raise an *AttributeError*.

Output

```
Employee has id 501 , name Ayaansh having phone number
89746827937
Employee has id 502 , name Rakesh
```

*10.3.1.1.3   Creating an Instance Variable Inside the Instance Method by
              Using the Self Variable*

We can also declare instance variables inside the instance method by using a self-variable. If any instance variable is declared inside the instance method, that instance variable will be added to the object once we call that method.

Example 10.15

```
class Employee:
  def __init__(self):
    self.id = 502
    self.name = "Srusti"
  def job(self):
    self.position = "Engineer"
Emp1= Employee()
Emp2= Employee()
Emp1.id = 501
Emp1.name = 'Nakshatra'
Emp1.job()
Emp2.job()
print("Employee has name ", Emp1.id,", is an ", Emp1.
position)
```

```
print("Employee has name ", Emp2.id,", is an ", Emp2.
position)
```

Observe that we have defined a method named *job()*. Inside that method we have defined one more variable *self.position*. In order to access the variable, we need to call the method. As *Emp1* and *Emp2* call this method, the *position* variable is added to both the objects.

Output

```
Employee has name 501 , is an Engineer
Employee has name 502 , is an Engineer
```

The following points need to be remembered regarding object variable:

- If a class has *n* objects, then there will be *n* separate copies of the object variable created as each object will have its own object variable.
- The object variable is not shared between objects.
- A change made to the object variable by one object will not be reflected in other objects.

10.3.1.2   Class Variables (Static Variables)

If the value of a variable is not varying from object to object, such variables need to be declared within the class but outside any method. Such types of variables are called *static variables* or *class variables*. These variables are tied to the class object they belong to and are independent of any class instances. For a class, only one copy of a static variable will be created and it is shared by all objects of that class.

We can create a static variable in five different ways:

- Creating static variables inside the class.
- Creating static variables inside the initializer by using class name.
- Creating static variables inside the instance methods by using class names.
- Creating static variables inside the static method by using class name.
- Creating static variables inside the class method by using either class name or *cls* variable.

10.3.1.2.1   Creating Static Variables Inside the Class

The static variable can be created inside the class but outside any method by simply writing the variable name and assigning some value to it.

Example 10.16

```
class Employee:
  WorkPlace= 'KaamDhaam'
    def __init__(self):
      self.id = 502
      self.name = 'Sushobhan Raj'
```

```
Emp1= Employee()
Emp2= Employee()
Emp1.id = 501
Emp1.name = 'Tusarika'
print("Employee: ", Emp1.name,", working at: ",Employee.
WorkPlace)
print("Employee: ", Emp2.name,", working at: ",Employee.
WorkPlace)
```

Observe that there is a class variable *WorkPlace* defined inside the class with the value *KaamDhaam* but it is not inside any method. We can access this class variable by *Employee.WorkPlace*. The general syntax for accessing a class variable is **class_name.variable_name**.

Output

```
Employee: Tusarika , working at: KaamDhaam
Employee: Sushobhan Raj , working at: KaamDhaam
```

10.3.1.2.2   Creating Static Variables Inside the Initializer by Using Class Name
The static variables can be created inside the *__init__()* method (initializer) by simply writing the *class_name.variable_name* and assigning some value to it.

Example 10.17

```
class Employee:
  def __init__(self):
    self.id= 502
    self.name= 'Ayaansh'
Employee.Workplace= "KaamDhaam"
Emp1 = Employee()
print("Employee:", Emp1.name,", Working at: ", Employee.
Workplace)
```

Observe that the class variable is defined inside the *__init__()* function simply by writing *Employee.WorkPlace= "KaamDhaam"*.

10.3.1.2.3   Creating Static Variables Inside the Instance Method by Using Class Name:
The static variables can be created inside any instance method of the class by simply writing *class_name.variable_name* and assigning some value to it. An instance method is a method, which has *self* as a parameter.

Example 10.18

```
class Employee:
  def __init__(self):
```

```
    self.id= 502
    self.name= "Ayaansh"
  def job(self):
    Employee.WorkPlace= "KaamDhaam"
Emp1= Employee()
Emp1.job()
print("Employee:", Emp1.name,", working at: ", Employee.
WorkPlace)
```

Observe that the *job(self)* method is an instance method. We have assigned the value "KaamDhaam" to *Employee.WorkPlace* inside the instance method.

10.3.1.2.4 Creating Static Variables Inside the Static Method by Using Class Name

The static variables can be created inside any static method of the class by simply writing *class_name.variable_name* and assigning some value to it. A static method is a method which does not have *self* as a parameter.

Example 10.19

```
class Employee:
  def __init__(self):
    self.id= 502
    self.name= " Ayaansh "
  def job():
    Employee.WorkPlace= "KaamDhaam"
Emp1 = Employee()
Employee.job()
print("Employee: ", Emp1.name, ", working at: ", Employee.
WorkPlace)
```

Observe that the *job()* method is a static method. We have assigned the value "KaamDhaam" to *Employee.WorkPlace* inside the static method.

10.3.1.2.5 Creating Static Variables Inside the Class Method by Using Either Class Name or cls Variable

The static variables can be created inside the class method of the class by simply writing *class_name.variable_name* and assigning some value to it. A method is a class method if we are using a decorator *@classmethod* or by passing a parameter *cls*.

Example 10.20

```
class Employee:
  def __init__(self):
    self.id = 502
    self.name = " Ayaansh "
```

```
@classmethod
def job(cls):
  Employee.WorkPlace= "KaamDhaam"

Emp1= Employee()
Employee.job()
print("Employee: ", Emp1.name,", Working at: ", Employee.
WorkPlace)
```

Observe that the *job()* method is a class method. We have assigned the value "KaamDhaam" to *Employee.WorkPlace* inside the class method.

We can access static variables by referencing the class name. For all the above examples, the output remains the same.

Output

```
Employee: Ayaansh , working at: KaamDhaam
```

The following points need to be remembered regarding class variable:

- If a class variable is defined in a class, then there will be one copy only for that variable. All the objects of that class will share the class variable.
- Since there exists a single copy of the class variable, any change made to the class variable by an object will be reflected in all other objects.

10.3.2   ACCESSING VARIABLES

It is not enough only to define different variables and assign values to them but we need to access them too. There are different methods for accessing these variables in the program.

The different methods for accessing variables are:

1. Accessing the variable by object reference.
2. Accessing the variable by class name.
3. Accessing the static variable outside the class by class name.

10.3.2.1   Accessing the Variable by Object Reference

We can access the variable by object reference. The general syntax to access a variable is: ***object_name.variable_name***.

Example 10.21

```
class Employee:
  WorkPlace= "KaamDhaam"
    def __init__(self):
      self.id= 502
      self.name = "Gauri"
    def job(self):
      self.position = "Manager"
```

```
Emp1= Employee()
Emp2= Employee()
Emp1.id = 501
Emp1.name = "Akarsh"
Emp1.job()
Emp2.job()
print("Employee: ", Emp1.name, ", working at: ", Emp1.
WorkPlace)
print("Employee: ", Emp2.name, ", working at: ", Emp2.
WorkPlace)
```

Observe that we have defined a class variable *WorkPlace*. In order to access the variable we can use the object reference that is *Emp1.WorkPlace* and *Emp2. WorkPlace*.

Output

```
Employee: Akarsh , working at: KaamDhaam
Employee: Gauri , working at: KaamDhaam
```

10.3.2.2   Accessing the Variable by Class Name

Instead of using the object name we can also use the class name while accessing the static variable. As the static variable is shared by the entire object created, we can use class name to access the static variable. The general syntax to access a static variable is: ***class_name.variable_name***.

Example 10.22

```
class Employee:
  WorkPlace= "KaamDhaam"
  def __init__(self):
    self.id = 502
    self.name= "Gauri"
Emp1= Employee()
Emp2= Employee()
Emp1.id= 501
Emp1.name= "Akarsh"
print("Employee: ", Emp1.name,", working at: ", Employee.
WorkPlace)
print("Employee: ", Emp2.name,", working at: ", Employee.
WorkPlace)
```

Observe that we have access to the same static variable by writing the *Employee. WorkPlace*.

Output

```
Employee: Akarsh , working at: KaamDhaam
Employee: Gauri , working at: KaamDhaam
```

10.3.2.3   Accessing the Static Variable Outside the Class by Class Name

The static variables can also be accessed outside the class. The way we assign values to a class variable, in the same way we can access it too. The general syntax to access a static variable outside the class is: ***class_name.variable_name***.

Example 10.23

```
class Employee:
  WorkPlace= "KaamDhaam"
  def __init__(self):
    self.id= 502
    self.name= "Gauri"
Emp1 = Employee()
print("Employee: ", Emp1.name,", working at: ", Employee.
WorkPlace)
Employee.WorkPlace= "KaamKaaj"
print("Employee: ", Emp1.name,", working at: ", Employee.
WorkPlace)
```

Observe that there is a class variable *WorkPlace* defined inside the class with value "KaamDhaam". This value is changed to "KaamKaaj" after the creation of the object. There may be several class variables defined for a class; we need to mention for which class variable the value is changed. It is also possible to define a class variable the same way we change the value of a class variable.

Output

```
Employee: Gauri , working at: KaamDhaam
Employee: Gauri , working at: KaamKaaj
```

10.4   PUBLIC, PRIVATE AND PROTECTED VARIABLES

We have seen that the data member can be an instance variable or static variable (class variable). We further categorize variables as public, protected and protected variables. These are also called *access specifiers.*

Public variables are variables that are defined in the class and can be accessed from anywhere in the program, of course by using the dot operator. Private variables, on the other hand, are variables that are defined in the class with a single or double underscore prefix. The private variables can be accessed only from within the class and from nowhere outside the class.

Unlike any other programming language Python does not have private, public and protected keywords, rather they are identified by no underscore, single or double underscore symbol before the variable/method.

10.4.1   Variable Starts Without Any _ (Underscore) Symbol

If a variable/method has no underscore before the variable/method name, it indicates that the name is public. So it is available to everyone and can be accessed from anywhere in the program.

Example 10.24

```
class Employee:
  def __init__(self,n,p):
    self.name = n
    self.phone = p
  def show(self):
    print("From show() method: ", self.name, self.phone)

Emp1= Employee("Rakesh", 9834681313)
Emp1.show()
print("Accessing data from outside the class: ")
print("Employee name ", Emp1.name)
print("Employee phone ", Emp1.phone)
print("Data saved by Emp1 Object: ", Emp1.__dict__)
```

Observe that there is no underscore preceding the variable name (*name* and *phone*). We can see that these variables can be accessed from within the class (*show()* method) and from outside the class (*Emp1.name* or *Emp1.phone*). The last print statement shows how the data is saved internally by the object *Emp1*.

Output

```
From show() method: Rakesh 9834681313
Accessing data from outside the class:
Employee name Rakesh
Employee phone 9834681313
Data saved by Emp1 Object: {'name': 'Rakesh', 'phone':
9834681313}
```

10.4.2 Variable Starts with _ (Single Underscore) Symbol

If a variable/method has an underscore before the name it indicates that it is for internal use only and it can be modified whenever the class wants to do so.

Python does not make these names truly private. So these names can be called directly from other modules. Such variables are also known as *weak private* or *protected*. We can use these names within the class, within the child class and within the package but not outside the package.

Example 10.25

```
class Employee:
  def __init__(self,n,p):
    self._name= n
    self._phone= p
  def show(self):
    print("From show() method: ", self._name, self._phone)

Emp1= Employee("Rakesh", 794534585)
Emp1.show()
```

```
print("Accessing Data from outside the class: ")
print("Employee name ", Emp1._name)
print("Employee phone", Emp1._phone)
print("Data saved by Emp1 object: ", Emp1.__dict__)
```

Observe the variable *self._name* and *self._phone,* have a single underscore (_) before the variable name. We can access this variable anywhere in the class or outside the class. Without using an underscore while accessing the name will raise an error *'AttributeError' : 'Class_Name' object has no attribute 'Attribute_Name'.*

Output

```
From show() method: Rakesh 794534585
Accessing Data from outside the class:
Employee name Rakesh
Employee phone 794534585
Data saved by Emp1 object: {'_name': 'Rakesh', '_phone': 794534585}
```

10.4.3   VARIABLE STARTS WITH _ _ (TWO UNDERSCORES) SYMBOLS

If a variable/method has a double underscore (__) before the name it indicates that it is for internal use only and it can be modified only inside the class. Such types of names are known as *strong private.*

We can use such variable names multiple times in different methods, as it is not possible to access private members directly from another class. The main purpose for double underscore (__) is to use names (variable/method) in class only if we do not want to use them outside of the class.

Example 10.26

```
class Employee:
  def __init__(self, n,p):
    self.__name= n
    self.__phone= p
  def show(self):
    print("From show() method: ", self.__name, self.__phone)

Emp1= Employee("Rakesh", 7854562555)
Emp1.show()
print("Data saved by Emp1 Object: ", Emp1.__dict__)
print("Accessing data from outside the class: ")
print("Employee name", Emp1.__name)
print("Employee phone", Emp1.__phone)
```

Observe the variables *self.__name* and *self.__phone.* Double underscore (__) is used before the variable name. These variables can be accessed only from within the class and nowhere from outside the class. As the variable names are private variables when we try to access them from outside the class it will raise an error *"AttributeError".*

Output

```
From show() method: Rakesh 7854562555
Data saved by Emp1 Object: {'_Employee__name': 'Rakesh',
'_Employee__phone': 7854562555}
Accessing data from outside the class:
Traceback (most recent call last):
File "C:\Users\RakeshN\AppData\Local\Programs\Python\Python3
7-32\chap-10.py", line 439, in <module>
print("Employee name", Emp1.__name)
AttributeError: 'Employee' object has no attribute '__name'
```

It is also possible to access these private variables but it is a bit tricky. We cannot access these variables the same way we have accessed the public and protected variables as discussed earlier.

Observe the second print statement in the previous example which has *Emp1.__dict__*. It displays the internal data structure along with data for the object *Emp1*. It is a dictionary (key–value pair) with keys *_Employee__name* and *_Employee__phone*. We can access the private data by using these keys which are explained in the example.

Example 10.27

```
class Employee:
  def __init__(self, n,p):
    self.__name= n
    self.__phone= p
  def show(self):
    print("From show() method:", self.__name, self.__phone)
Emp1= Employee("Rakesh", 7854558565)
Emp1.show()
print("Data saved by Emp1 Object:", Emp1.__dict__)
print("Accessing data from outside the class: ")
print("Employee name", Emp1._Employee__name)
print("Employee phone", Emp1._Employee__phone)
```

Observe that the variables name and phone are prefixed with two underscores indicating they are private variables. In the last two print statements the data stored in these variables are printed but we have used the dictionary keys to access them.

Output

```
From show() method: Rakesh 7854558565
Data saved by Emp1 Object: {'_Employee__name': 'Rakesh',
'_Employee__phone': 7854558565}
Accessing data from outside the class:
Employee name Rakesh
Employee phone 7854558565
```

10.5   METHODS

We have discussed that a function in object-oriented concept is known as a method. A method is a function that is associated with a particular class. Methods are defined inside a class. This definition of a method makes the relationship between the class and the method explicit. The syntax for invoking a method is different from the syntax for calling a function.

There are three different types of methods:

1. Instance method.
2. Class method.
3. Static method.

10.5.1   Instance Method

A method that deals with the instance variable is known as *instance method*. The instance method must have a *self* as an argument. Apart from *self*, we may/may not have any other arguments.

The first argument is always *self* in the parameter list. Moreover, if the method only contains *self* as a single parameter, we do not pass any value for this parameter while calling the method. Python provides the value *self* automatically. The self-argument refers to the object itself, the object that has called the method. It means, if a method that takes no arguments, it should accept *self*. Similarly, a method defined to accept one parameter will actually take two parameters, *self* and the *parameter*.

Example 10.28

```
class Employee:
  def __init__(self, salary):
    self.salary = salary

  def Set_Value(self):
    self.id = 500
    self.name = "Rakesh"

  def read_Value(self, i , n):
    self.id = i
    self.name = n

Emp1 = Employee(5000)
Emp1.Set_Value()
print(Emp1.id," Employee Name ", Emp1.name, " with salary ",
Emp1.salary)

Emp1.read_Value(501,"Naveen")
print(Emp1.id, " Employee Name ", Emp1.name, " with
salary ", Emp1.salary)
```

Observe that we have defined three methods namely __init__(), Set_Value() and read_Value(). The first one is an initializer with two arguments *self* and *salary*. In the second one, we have only one argument that is *self*. In the third one, the arguments are *self*, *i* and *n*, which have the value that we passed through the parameters. In all cases, *self* is a part of the argument list so these are *instance methods*.

Output

```
500 Employee Name Rakesh with salary 5000
501 Employee Name Naveen with salary 5000
```

10.5.2 CLASS METHOD

If we want to deal with the class variable, we need a *class method*. Class methods are different from ordinary methods in two ways. First, they are called by a class (not by the instance of the class). Second, the first parameter of the class method is *cls* and not the *self*. Generally, class methods are used when we instantiate a class, using different parameters than those usually passed to the class constructor. In the case of variables, class variables are the same as static variables but it is different in the case of methods.

Example 10.29

```
class Employee:
    organization = "KaamDhaam"

  def Set_Value(self):
    self.id= 502
    self.name= "Rakesh"

  def info(cls):
    return cls.organization

Emp1= Employee()
Emp1.Set_Value()
print(Emp1.id," Employee Name ", Emp1.name," in ", Employee.
info(Employee), " Organization" )
```

Observe that a class variable *organization* is defined. To access this class variable, we need a class method. Here the method *info(cls)* is a class method as it takes *cls* as a parameter. We can see that to access the returned value of this class method, we have written *Employee.info(Employee)*. It means, in the Employee class, there is a class method *info(cls)*, we need to pass the class name *Employee* as a parameter.

Observe the next example where in order to avoid passing the class name as a parameter we can use a decorator. We can write *@classmethod* decorator before the class method. In this case, we need not pass the class name as a parameter while calling a class method. We have already discussed decorators in section 8.11.

Example 10.30

```
class Employee:
    organization = "KaamDhaam"
  def Set_Value(self):
    self.id = 502
    self.name = "Rakesh"

  @classmethod
  def info(cls):
    return cls.organization

Emp1 = Employee()
Emp1.Set_Value()
print(Emp1.id," Employee Name ", Emp1.name," in ", Employee.
info(), " Organization")
```

Observe the last print statement, we have not passed any parameter to *Employee.info()*. In either case the output remains the same.

Output

```
502 Employee Name Rakesh in KaamDhaam Organization
```

10.5.3 STATIC METHOD

If we want a method that has nothing to do either with the instance variable or class variable, we have with us what are known as *static methods*. This method is useful when we want to perform any operation with another class or object. Any functionality that belongs to a class, but the objects do not require them, is the *static method*. Static methods are similar to class methods; the only difference is that a static method does not receive any additional arguments. They are just like normal methods that belong to a class.

We can make a method static by using a decorator *@staticmethod* and the static method is accessed by *class_name.method_name()*.

Example 10.31

```
class Employee:
  organization = "KaamDhaam Pvt. Ltd."
  def Set_Value(self):
    self.id = 502
    self.name = "Rakesh"

  @classmethod
  def info(cls):
    return cls.organization
```

```
@staticmethod
def print_info():
  print("This is a printed for static method")

Emp1 = Employee()
Emp1.Set_Value()
print(Emp1.id," Employee Name ", Emp1.name)
Employee.print_info()
```

Observe that we have a method *print_info()*, this method is a static method as we have used the decorator *@staticmethod*.

Output

```
502 Employee Name Rakesh
This is a printed for static method
```

10.6   CLASS INSIDE A CLASS (INNER CLASS)

Sometimes it is necessary to define a class inside another class. Such a class is known as *inner class*. In order to define an inner class, it is evident that we must have an outer class. For example, every college has departments. Without the existence of the college class, there is no existence of the department class. So the department class has to be the inner class of the class college. We can create the object of the inner class in two ways:

1. Object of the inner class is created inside the outer class.
2. Object of the inner class is created outside the outer class.

In order to understand this let us take an example, Let us assume a laptop is given to each employee of an organization. Each laptop is different from the other in terms of Physical address, RAM and CPU. So laptop class is defined inside employee class.

10.6.1   OBJECT OF THE INNER CLASS IS CREATED INSIDE THE OUTER CLASS

The object of the inner class is created inside the outer class by *self. InnerClassObjectName = self.InnerClassName()*.

Example 10.32

```
class Employee:

  def __init__(self,sal):
    self.salary= sal
    self.lap= self.laptop()
```

```
    def show(self):
      print(self.salary)

  class laptop:
    def __init__(self):
      self.make= 'Dell'
      self.ram= '4Gb'
      self.cpu= 'Intel i7'
```

Observe that we have defined a class *laptop* inside the class *Employee*. The statement *'self.lap = self.laptop()'* inside the *__init__()* created an object of inner class (laptop) in the outer class (Employee). This statement is inside the constructor of the *Employee* class. When the object of the *employee* is created, the object of the *laptop* will also get created.

10.6.2   OBJECT OF THE INNER CLASS IS CREATED OUTSIDE THE OUTER CLASS

The object of the inner class is created outside the outer class by *InnerClassObject Name = OuterClassName.InnerClassName()*.

Example 10.33

```
class Employee:
  def __init__(self, sal):
    self.salary = sal

  def show(self):
    print(self.salary)

  class laptop:
    def __init__(self):
      self.make = "Dell"
      self.ram = '4gb'
      self.cpu = "AMD RYZEN"

Emp1 = Employee(50000)
Emp1.show()
lap1 = Employee.laptop()
print(lap1.make, lap1.cpu, lap1.ram)
```

The class *laptop* is defined inside the class *Employee*. Observe the statement *lap1 = Employee.laptop()*. This statement creates an object *lap1* which is an instance of class *laptop*.

10.6.3   ACCESSING ATTRIBUTES OF INNER CLASS

Accessing a name defined inside the inner method is a bit tricky. The different ways of accessing the attributes of inner class are:

1. Accessing the attributes of inner class outside the class
2. Accessing the attributes of inner class in the outer class
3. Accessing the attributes of inner class outside the outer class
4. Accessing the attributes of inner class when inner class and outer class have the same method

10.6.3.1 Accessing the Attributes of Inner Class Outside the Class

In order to access the attribute of the inner class outside the class, the general syntax is: ***ObjectofOuterClass.ObjectofInnerClass.AttributeNameofInnerObject***.

Example 10.34

```
Emp1 = Employee(50000)
Emp1.show()
print(Emp1.lap.make, Emp1.lap.cpu, Emp1.lap.ram)
```

This code is in continuation with example 10.32. In example 10.32, observe that the object of inner class is created inside the outer class. The last line, print the data of inner class in the main program.

10.6.3.2 Accessing the Attributes of Inner Class in the Outer Class

In order to access the attribute of the inner class outside the outer class, the general syntax is: ***self.ObjectofInnerClass.AttributeNameofInnerObject***.

Example 10.35

```
class Employee:
  def __init__(self,sal):
    self.salary = sal
    self.lap = self.laptop()

  def show(self):
    print(self.salary)
    print(self.lap.make, self.lap.cpu, self.lap.ram)

  class laptop:
    def __init__(self):
      self.make = "DELL"
      self.ram = "4gb"
      self.cpu = "i7"

Emp1 = Employee(50000)
Emp1.show()
```

Observe the *show()* method. This method is defined in the *Employee* class. See the last line in the *show()* method, "*self.lap.make*" can access the attribute *make*, which is an attribute of inner class.

10.6.3.3   Accessing the Attributes of Inner Class Outside the Outer Class

We have already seen that an object of inner class can be created outside the outer class. In order to access the attribute of the inner class outside the outer class, the general syntax is: ***ObjectofInnerClass.AttributeNameofInnerObject***.

Example 10.36

```python
class Employee:
   def __init__(self,sal):
      self.salary = sal

   def show(self):
      print(self.salary)

   class laptop:
      def __init__(self):
         self.make= "DELL"
         self.ram= "4Gb"
         self.cpu= "i7"

Emp1 = Employee(50000)
Emp1.show()
lap1= Employee.laptop()
print(lap1.make, lap1.cpu, lap1.ram)
```

10.6.3.4   Accessing the Attributes of Inner Class When Inner Class and Outer Class Have the Same Method

If a method has the same name in both inner and outer class, we can call the inner class by: ***self.innerClassObjectName.innerClassMethod()***.

The example below shows that a function named *show()* is defined in both inner class and outer class. In order to access the method of inner class in the outer class, we have *self.lap.show()*. Whenever the *show()* method of the outer class is called, it automatically calls the *show()* method of inner class.

Example 10.37

```python
class Employee:

   def __init__(self, sal):
      self.salary = sal
      self.lap = self.laptop()

   def show(self):
      print(self.salary)
      self.lap.show()

   class laptop:
      def __init__(self):
```

```
        self.make = "Dell"
        self.ram = "4gb"
        self.cpu = "i7"

    def show(self):
        print(self.make,self.cpu,self.ram)
Emp1 = Employee(50000)
Emp1.show()
```

All the examples above give the same output.

Output

```
50000
Dell i7 4gb
```

10.7   SOME SPECIAL METHODS AND ATTRIBUTES

When Python objects are created, a small set of predefined properties and methods are also available with it. These methods are also known as *magic methods* that are commonly used for operator overloading. Observe that these methods are preceded and followed by two underscores (__). These methods can be listed out by *print(dir(class_name))*. These functions can be used freely as if they are defined in the class.

Example 10.38

```
class Employee:
    pass
print(dir(Employee))
```

Output

```
['__class__', '__delattr__', '__dict__', '__dir__', '__
doc__', '__eq__', '__format__', '__ge__', '__
getattribute__', '__gt__', '__hash__', '__init__',
'__init_subclass__', '__le__', '__lt__', '__module__',
'__ne__', '__new__', '__reduce__', '__reduce_ex__', '__
repr__', '__setattr__', '__sizeof__', '__str__', '__
subclasshook__', '__weakref__']
```

Let us discuss some of these methods.

10.7.1   _ _ DICT _ _ ()

The method *__dict__()* (it is a dictionary) returns the names and values of all the properties (variables) the object is currently carrying. The general syntax is: *object_name.__dict__*.

Example 10.39

```
class Employee:
  def __init__(self):
    self.id= 502
    self.name= "Rakesh"

Emp1 = Employee()
print(Emp1.__dict__)
```

Observe the last line *print(Emp1.__dict__)*. It returned all the attributes and their values which are present in the object *Emp1*. As a class may contain the variables but not any values so it is natural that this method is not applicable to class.

Output

```
{'id': 502, 'name': 'Rakesh'}
```

10.7.2 _ _ NAME _ _ ()

This function returns *__main__()*, if the python interpreter running the program as main program. It means the module is running standalone. Otherwise it will return the name of the module if the file is imported from another module.

Example 10.40

```
def my_fun1():
  print("The name of the module is: ",__name__)

if __name__ == '__main__':
  my_fun1()
```

When we are trying to print *__name__* it returned *__main__*, as it is a standalone program.

Output

```
The name of the module is: __main__
```

But when we try to find the *__name__* while calling it from the module, it returns the module name. In the example below, we have a module named *my_module* and inside that we have a method *My_method()* that prints *__name__*.

Example 10.41

```
def My_method():
  print("Module name ", __name__)
```

Now we have imported "*my _ module*" and *math*.

```
import my_module
import math
  if __name__ == "__main__":
    my_module.My_method()
    print(math.__name__)
```

Observe that we called *My_method()* which is defined inside *my_module* and in the last line we printed *math.__name__*. The output shows the module name of both the modules.

Output

```
Module name my_module
math
```

10.7.3 _ _ DOC _ _

This function returns the documents available with the class/object/module. The general syntax is: ***Class_Name[/ Object_Name / Module_Name].__doc__***

The documentation is nothing but the String written for better understanding. These are the strings which are written in the class/object/module with a pair of triple quotes.

Example 10.42

```
import math
class A:
  '''This is a simple Class Document.'''
  pass

Obj_A= A()
print("\n---for class---")
print(A.__doc__)
print("\n---for Object---")
print(Obj_A.__doc__)
print("\n---for Math Module---")
print(math.__doc__)
```

In order to create the object the class is called so the output for *A.__doc__* and *Obj_A.__doc__* has the same output. The last print statement displays the document written inside the math module.

Output

```
---for class---
This is a simple Class Document.
```

```
---for Object---
This is a simple Class Document.

---for Math Module---
```

This module provides access to the mathematical functions
defined by the C standard.

10.7.4　HASATTR()

Python provides a function which checks if an object contains a specified property. If
an attribute is available in the given object, it returns True, otherwise it returns False.
The general syntax is: *hasattr(object_name, property)*. This function expects two
arguments, the object being checked; and the name of the property whose existence
has to be reported. The attribute name needs to be passed as a String.

Example 10.43

```
class Employee:

  def __init__(self):
    self.id = 502
    self.name = "Rakesh"

  @classmethod
  def job(cls):
    Employee.WorkPlace= "xxxyyyzzz"

Emp1= Employee()
Employee.job()
if hasattr(Emp1, "id"):
  print("id is a property of the object.")
else:
  print("id is not a property of the object.")

if hasattr(Employee,"id"):
  print("id is a property of the class.")
else:
  print("id is not a property of the class.")
```

Output

```
id is a property of the object.
id is not a property of the class.
```

10.7.5　GETATTR()

This function is used to access the attribute of an object. The general syntax is:
getattr(object_name, attribute_name [,default]).

As *getattr()* is a built-in function and not a method, it is not called using the dot operator. This function takes three parameters. The first parameter is the object itself. The second parameter is the name of the attribute as a String, and the optional third parameter is the default value to be returned if the attribute does not exist. This function returns the value of the attribute if the variable exists. If the attribute name does not exist in the object's namespace and the default value is also not specified, then an exception will be raised. If the optional third parameter (a String, an Integer, a Boolean value or a real number) is defined, it will return the third parameter if the attribute is not a part of the object.

Example 10.44

```
class Employee:
  def __init__(self):
    self.id= 502
    self.name= "Ayaansh"

Emp1= Employee()
a = getattr(Emp1, "id", " an attribute of the object")
print(a)
b= getattr(Emp1,"salary", " not an attribute of the object.")
print(b)
```

Observe that *Emp1* has attributes *"id"* and *""*. When we applied *getattr* on *"id"*, it returned the value of the attribute and when we applied *getattr* on *salary*, it returned the String defined in the third parameter.

Output

```
502
not an attribute of the object.
```

10.7.6   SETATTR(OBJ,NAME,VALUE)

This function is used to set an attribute of the object. If the attribute does not exist, then the attribute will be created and assign the specified value to it. This new attribute is available to a specific object only. The general syntax is: *setattr(obj_name,attribute_name,value)*. It takes three parameters. The first parameter to *setattr()* function is the *object*, the second parameter is the name of the *attribute* and the third is the value of the specified attribute.

Example 10.45

```
class Employee:
  def __init__(self):
    self.id = 502
    self.name = "Rakesh"
```

```
Emp1 = Employee()
setattr(Emp1,"gender","Male")
print(Emp1.id, Emp1.name, Emp1.gender)
```

Observe that we have created a new attribute *"gender"* for the object Emp1. And assigned the value *"Male"* to it. A new attribute is created and a value is assigned to it and if the attribute already exists, a new value is assigned to it.

Output

```
502 Rakesh Male
```

When the attribute is created by using the *setattr* method, it is available only to that specific object in which it is defined. If we want to access the attribute gender from any other object, it will show *AttributeError*.

Example 10.46

```
class Employee:
  def __init__(self):
    self.id= 502
    self.name= "Sushobhan Raj"

Emp1 = Employee()
setattr(Emp1,"gender","Male")
print(Emp1.id, Emp1.name, Emp1.gender)

Emp2= Employee()
print(Emp2.id, Emp2.name, Emp2.gender)
```

Observe that *setattr* is used for an *Emp1* object that creates a new attribute *gender* with value *Male*. When we want to access the gender attribute of *Emp2*, it raises an *AttributeError*.

Output

```
502 Sushobhan Raj Male
Traceback (most recent call last):
File "C:\Users\RakeshN\AppData\Local\Programs\Python\Python3
7-32\chap-10.py", line 880, in <module>
print(Emp2.id, Emp2.name, Emp2.gender)
AttributeError: 'Employee' object has no attribute 'gender'
```

10.7.7 DELATTR()

This function deletes an existing attribute. Once deleted, the attribute no longer belongs to the object. The general syntax is: *delattr(obj,name)*. It takes two parameters, the first parameter is the object name and the second parameter is the name of the attribute.

Example 10.47

```
class Employee:
  def __init__(self, i,n,g):
    self.id=i
    self.name= n
    self.gender= g

Emp1= Employee(501,"Naveen","Male")
Emp2= Employee(502,"Rakesh","Male")

print("----Before delattr()")
print(Emp1.id, Emp1.name, Emp1.gender)
print(Emp2.id, Emp2.name, Emp2.gender)

print("----After delattr()")
delattr(Emp2, "id")
print(Emp1.id, Emp1.name, Emp1.gender)
print(Emp2.id, Emp2.name, Emp2.gender)
```

Observe in the above code that there are two objects (Emp1 and Emp2) with attributes *id, name* and *gender* having value (*501,"Naveen","Male"*) and (*502,"Rakesh","Male"*) respectively. Calling *delattr(Emp2,"id")* results in deletion of the attribute *id* from object *Emp2*.

After the *delattr()* method is called the attributes of object *Emp1* remain intact whereas for object *Emp2* an error *AttributeError* message appears. It means *delattr()* method deletes attribute of the object passed as argument and not attribute of other objects.

Output

```
----Before delattr()
501 Naveen Male
502 Rakesh Male
----After delattr()
501 Naveen Male
Traceback (most recent call last):
  File "C:\Users\RakeshN\AppData\Local\Programs\Python\Py
  thon37-32\chap-10.py", line 908, in <module>
    print(Emp2.id, Emp2.name, Emp2.gender)
AttributeError: 'Employee' object has no attribute 'id'
```

10.7.8   DEL

This function deletes the static variable. The general syntax is: *del class_name.variable_name*. The *delattr()* function deletes attribute of an object whereas *del* deletes a class attribute.

Example 10.48

```
class Employee:
  WorkPlace = "KaamDhaam Pvt. Ltd."
    def __init__(self):
      self.id= 502
      self.name= "Rakesh"

Emp1 = Employee()
print(Emp1.id, "works at", Emp1.WorkPlace)

del Employee.WorkPlace
print(Emp1.id, "works at ", Emp1.WorkPlace)
```

Observe that *WorkPlace* is a class attribute (static variable) with value "KaamDhaam Pvt. Ltd.". Before calling the *del* function the output is *502 works at KaamDhaam Pvt. Ltd.* but after calling the *del* function the output is an error message *AttributeError*. It means the variable *WorkPlace* does not exist anymore.

Output

```
502 works at KaamDhaam Pvt. Ltd.
Traceback (most recent call last):
  File "C:\Users\RakeshN\AppData\Local\Programs\Python\Py
  thon37-32\chap-10.py", line 929, in <module>
    print(Emp1.id, "works at ", Emp1.WorkPlace)
AttributeError: 'Employee' object has no attribute 'WorkPlace'
```

10.7.9   ISINSTANCE()

This method returns True, if an object is of specific class(type) otherwise it returns False. The general syntax is: *isinstance(Object_name, Class_Name)*. This method requires two parameters, the first one is the name of the object and the second parameter is the name of the class.

Example 10.49

```
class A:
  pass
class B:
  pass

Obj_A = A()
  x= isinstance(Obj_A, A)
print(x)

Obj_B = B()
y = isinstance(Obj_B, A)
print(y)
```

Here two classes (A and B) are defined and two objects *Obj_A* and *Obj_B* are created. Observe that *isinstance(Obj_A, A)* returned *True* and *instance(Obj_B, A)* returned *False*. If we are trying to take an object which is never created *isinstance(Obj_C, A)* it will raise a *NameError*.

Output

```
True
False
```

10.7.10   REPR()

This function returns a String representation of an object. The general syntax is *repr(object)*. The function works on any object, not just class instances.

Example 10.50

```
class Employee:

def __init__(self, i,n,g):
  self.id = i
  self.name = n
  self.gender = g

Emp1 = Employee(501,"Naveen","Male")
Emp2 = Employee(502, "Rakesh", "Male")
print(repr(Emp1))
```

Observe that *repr(Emp1)* returned the address of *Emp1* which is an object of class *Employee*.

Output

```
<__main__.Employee object at 0x03B75D60>
```

WORKED OUT EXAMPLES

Example 10.51 Write a program to create a class, initialize with the length and breadth of the rectangle. Define two methods that will find the area and perimeter of the rectangle.

```
class rect:
  def __init__(self, l,b):
    self.length = l
    self.breadth = b

  def area(self):
    return self.length * self.breadth
```

```
    def perimeter(self):
        return 2 * (self.length + self.breadth)

leng = int(input("Enter length of rectangle: "))
br = int(input("Enter breadth of rectangle: "))

obj_rect = rect(leng,br)
print("The area is ", obj_rect.area())
print("The perimeter is", obj_rect.perimeter())
```

Output

```
Enter length of rectangle: 2
Enter breadth of rectangle: 3
The area is 6
The perimeter is 10
```

Example 10.52 Write a program to find summation of two times. Create a class, define two methods that will add two times and show the new time.

```
class time():
    def __init__(self,h,m):
        self.hours = h
        self.mins = m

    def TimeAddition(t1,t2):
        t3 = time(0,0)
        if (t1.mins + t2.mins >= 60):
        t3.hours = (t1.mins + t2.mins)//60
            t3.hours = t3.hours + t1.hours + t2.hours
            t3.mins = t1.mins + t2.mins - (((t1.mins + t2.mins)//60)*60)
            return t3

    def TimeShow(self):
        print(self.hours, "Hours and ", self.mins, " minutes")

T1 = time(2,54)
T2 = time(1,44)
T3 = time.TimeAddition(T1,T2)
T3.TimeShow()
```

Output

```
4 Hours and 38 minutes
```

Example 10.53 Write a program that defines a class Nationality and a static method print Nationality which prints the nation.

```
class Nationality(object):

  @staticmethod
  def printNationality(str):
    if (str == "American"):
      print("America")
    elif (str == "Indian"):
      print("India")

Native = Nationality()
Native.printNationality("Indian")
Nationality.printNationality("Indian")
```

Output

```
India
India
```

MULTIPLE CHOICE QUESTIONS

1. What is the output of the given code?
   ```
   class A:
       def __init__(self,a = 'Hi'):
           self.a = a
       def show(self):
           print(self.a)
   B = A()
   B.show()
   ```

   a. Hi
   b. 'Hi'
   c. "Hi"
   d. None of the above

2. What is the output of the given code?
   ```
   class A:
   def __init__(self,a,b,c):
   self.a = a + b * c
   a1 = A(1,2,3)
   a2 = getattr(a1,'a')
   setattr(a1, 'a',a2+1)
   print(a1.a)
   ```

   a. 7
   b. 8
   c. 9
   d. 10

3. What is the output of the given code?
```
class A:
    def __init__(self,a):
        self.a = a
    def show():
    print(self.a)
    a1 = A(5)
    a1.show()
```

   a. 5
   b. 0
   c. TypeError
   d. NameError

4. What is the output of the given code?
```
class A:
    def __init__(self):
        self.variable = 8
        self.Change(self.variable)
    def Change(self, var):
        var = 4
obj = A()
```

   a. 8
   b. 4
   c. TypeError
   d. AttributeError

5. What is the output of the given code?
```
class A:
    def __init__(self,a):
        self.price = a
obj = A(50)
obj.quantity = 10
obj.bags = 2
print(len(obj.__dict__))
```

   a. 50
   b. 10
   c. 2
   d. 3

6. Which of the following is not a class method?
   a. Static
   b. Non-Static
   c. Bounded
   d. Unbounded

7. Which of the following is required to create a new instance?
   a. A Class
   b. A Method
   c. A return statement
   d. An Attribute

8. What is the output of the given code?
   ```
   class A:
       def __init__(self,a):
           self.a = a
   a = 1
   value = A(555)
   print(value.a)
   ```
   a. 555
   b. 1
   c. Error
   d. None of the above

9. What of the following checks whether an object B is an instance of a class A?
   a. B. isinstance(A)
   b. A. isinstance(B)
   c. isinstance(A,B)
   d. isinstance(B,A)

10. The method which begins and ends with two underscore is called
    a. User-defined method
    b. Built-in Methods
    c. Special Methods
    d. Magic Methods

11. Which of the following is private data?
    a. __a__
    b. __a
    c. a__
    d. None of the above

12. Which is the work of getarrt()?
    a. Access attribute of an object
    b. Check for an attribute
    c. Set an attribute
    d. None of the above

13. Which of the following does __doc__ return ?
    a. strings available in a class
    b. document available in a class
    c. All the comment statements
    d. None of the above

14. Which of the following is used to create an empty class?
    a. class A:
       return
    b. class A:
       pass
    c. class A:
    d. None of the above

15. What is the work of hasattr()?
    a. To set an attribute
    b. To access the attribute of the object
    c. To delete an attribute
    d. To check if an attribute exists or not

16. What is the work of delattr()?
    a. To set an attribute
    b. To access the attribute of the object
    c. To delete an attribute
    d. To check if an attribute exists or not

17. What does __name__ return?
    a. Name of the file
    b. Name of the class
    c. Name of the function
    d. None of the above

18. A class name begins with
    a. class
    b. def
    c. return
    d. None of the above

19. Which of the following statements is not true?
    a. An object may contain other objects
    b. A reference variable refers to an object
    c. A reference variable is an object
    d. An object can contain the references to other objects

20. Which of the following statements is correct about inner class?
    a. An object may contain other objects
    b. A reference variable refers to an object
    c. A class may contain another class
    d. An class can contain the references to other class

DESCRIPTIVE QUESTIONS

1. What is a class? How do you define a class in Python?
2. What are the different types of attributes available in OOPL? What are the different ways of accessing these attributes?
3. What is an access specifier in OOPL? Describe with an example.
4. What are the different types of methods available in OOPL? What are the different ways of accessing these methods?
5. What do you mean by inner class? Describe the ways of accessing the attributes of an inner class.

PROGRAMMING QUESTIONS

1. Write a program to create a class, initialize with the temperature. Define two methods that will convert for Celsius to Fahrenheit and vice-versa.
2. Write a program that finds the power set of a given set. A power set is a collection of all unique subsets.
3. Write a program to solve a classic ancient Chinese puzzle: We count 3 heads and 10 legs among the birds and animals in a farm. How many birds and animals do we have?
4. Write a program that accepts a list of data, and a target, it should return a pair of numbers which sum equals target.

ANSWER TO MULTIPLE CHOICE QUESTIONS

1.	A	2.	B	3.	C	4.	D	5.	D
6.	B	7.	A	8.	A	9.	C	10.	C
11.	B	12.	A	13.	C	14.	B	15.	D
16.	C	17.	D	18.	A	19.	C,D	20.	C

11 Inheritance

LEARNING OBJECTIVES

After studying this chapter, the reader will be able to:

- Know different types of inheritance
- Use initialization in inheritance
- Understand the method resolution order
- Use some specialized methods

11.1 INTRODUCTION

In this chapter we are going to discuss inheritance. This is one among many important concepts of object-oriented programming (OOP) language. Let us try to understand this from real-life examples. Throughout our life, whatever property we accumulate, all of it is ours. Along with that, whatever our parents accumulated, that also, we can claim to be ours. The same concept can be implemented in programming languages which follow OOP concepts. Let us talk about doctors. An ophthalmologist is a doctor who specializes in the treatment of eyes. Whatever a general doctor can do, an ophthalmologist can also do, and apart from that he can also treat any problem related to the eyes. But it is not vice-versa. Here, we are talking about inheritance in doctors. Try to understand this concept with the help of an example presented in figure 11.1.

It's possible to build more specialized (more concrete) classes using some sets of predefined general rules and behaviors. When we want to implement a concept of *General-Special* or *is a* concept we need to implement inheritance. In the employee class, a manager is an employee. The manager has all the attributes and methods that an employee has. Along with all those attributes and methods, the manager performs some other tasks also. So, a manager can have some more attributes and methods. It means a manager inherits all of the attributes and methods of an employee. So, we can say a manager is an employee.

The class from which other classes are derived is known as *Super Class* or *Base Class*, *Parent Class* or *General Class*. A class which derives from another class is known as *Sub Class* or *Derived Class*, *Child Class* or *Special Class*.

Inheritance is a way of building a new class, not from scratch, but by using an already defined class. The new class inherits (and this is the key) all the already existing attributes and methods, and is able to add some new ones if needed. The advantage of using inheritance is code reusability, it provides a transitive property, also, it resembles real-life relationships between classes.

DOI: 10.1201/9781003219125-13

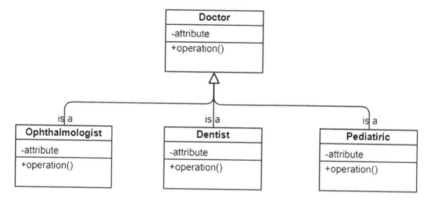

FIGURE 11.1 Example of "is a" relationship

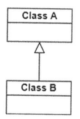

FIGURE 11.2 Single level inheritance

The inheritance can be of different types:

1. Single-Level Inheritance.
2. Multi-Level Inheritance.
3. Multiple Inheritance.
4. Hierarchical Inheritance
5. Hybrid Inheritance.

11.1.1 SINGLE-LEVEL INHERITANCE

Inheritance in which there is only one super class and one sub class, is called *single level inheritance*. In this inheritance, the child class can inherit all the properties of the parent class.

When Class B is a special class which inherits from a general Class A it can be represented by a directed hollow arrow from Class B to Class A as shown in figure 11.2.

Here "*Class A*" is the *Base-Class* and "*Class B*" is the *Derived-Class.*

The general syntax when Class B inherits form Class A is:

class A:
    Class_Suit of A
    :::
class B(A):
    Class_Suit of B
    :::

In inheritance notice that:

- The class to be inherited needs to be defined first.
- This class is defined like any normal class.
- The definition of the class that will inherit is the same as any normal class except that the super class is passed as a parameter.
- The child class can access all the properties of the parent class.

Example 11.1 Inheritance

```
class Employee:
  def EmpCanDo(self):
    print("Employee can do......")

class Manager(Employee):
  def MgrCanDo(self):
    print("Manager can do....")

print("....Employee Object Created.....")
Emp1 = Employee()
Emp1.EmpCanDo()

print("\n...Manager Object Created...")
Mgr1 = Manager()
Mgr1.MgrCanDo()
Mgr1.EmpCanDo()
```

In this example, we have a class name *Employee*, and a method *EmpCanDo()* defined inside the Employee class. We have another class *Manager* with a method *MgrCanDo()*. Observe that while creating the class Employee, there is no parameter passed but while creating the class *Manager*, a parameter *Employee* is passed. This indicates *Employee* is the *super class* (or *parent class*) and *Manager* is a *sub class* (or *child class*).

In other words, *Manager* is a *sub class* of the *Employee* class. When we create an object *Mgr1* of the class *Manager* it can call the method defined inside the *Manager* class as well as the methods defined inside the *Employee* class. This is because the *Manager* class inherits from the *Employee* class. It is evident from the output.

Output

```
...Employee Object Created...
Employee can do...

...Manager Object Created...
Manager can do...
Employee can do...
```

11.1.2 MULTI-LEVEL INHERITANCE

The concept of single-level inheritance can be extended to multi-level inheritance. When inheritance has the *child class*, *parent class* and *grandparent class*, where the child class inherits from the parent class, and the parent class inherits from the grandparent class, we call it *multi-level inheritance*.

In the figure 11.3, Class C is the child class, which inherits from Class B. So, there is a hollow arrow line from Class C to Class B. Here Class B is the parent class of Class C. Again, Class B inherits from Class A. So, there is a hollow arrow line from Class B to Class A. Class A is the grandparent Class of Class C. In multi-level inheritance, the child class inherits all the properties of its parent class as well as from its grandparent class.

The general syntax for multi-level inheritance is:

class A:
    Class_Suit of A
    :::
class B(A)
    Class_Suit of B
    :::
class C(B)
    Class_Suit of C
    :::

In multi-level inheritance notice that:

- It is an application of *single-level inheritance multiple times.*
- The class to be inherited need to be defined first.
- This class is defined like any normal class.
- The *parent class of a parent class* is known as *grandparent class.*
- The definition of the class that will inherit is the same as any normal class except super class is passed as a parameter.
- The *child class* can access all the properties of *parent class* as well as *grandparent class.*

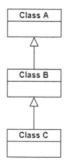

FIGURE 11.3 Multi-level inheritance

Assume that there is a class named *GeneralManager*. The *GeneralManager* is also a *Manager*. So, it must inherit from the class *Manager*. A Manager is an *Employee*. So, a *Manager* class inherits from *Employee* class. *Employee* class is the parent class and *Manager* class and *Manager* class is the parent of *GeneralManager* class. So, this inheritance is a *multi-level inheritance*.

Example 11.2 Multi-level inheritance

```
class Employee:
  def EmpCanDo(self):
    print("Employee can do...")

class Manager(Employee):
  def MgrCanDo(self):
    print("Manager can do...")

class GeneralManager(Manager):
  def GenMgrCanDo(self):
    print("General Manager can do...")

print("\n...Employee Object Created...")
Emp1 = Employee()
Emp1.EmpCanDo()

print("\n...Manager Object Created...")
Mgr1 = Manager()
Mgr1.MgrCanDo()
Mgr1.EmpCanDo()

print("\n...General Manager Object Created...")
GM1 = GeneralManager()
GM1.GenMgrCanDo()
GM1.MgrCanDo()
GM1.EmpCanDo()
```

Observe here that the *Employee* class is the *grandparent* class, *Manager* is the parent class and *GeneralManager* is the child class. When the object of *GeneralManager* is created, it can access the methods defined under *Manager* Class as well as the *Employee* class.

Output

```
...Employee Object Created...
Employee can do...

...Manager Object Created...
Manager can do...
Employee can do...
```

```
...General Manager Object Created...
General Manager can do...
Manager can do...
Employee can do...
```

11.1.3 Multiple Inheritance

It may happen that a class may inherit from two different classes. In other words, a child class has two different parent classes. This type of inheritance is called *multiple inheritance*.

In figure 11.4, Class C is the child class, which inherits from Class A and Class B. So, there is a hollow arrow line from Class C to Class B and Class C to Class A. Here Class A and Class B are the parent class and Class C is the child class.

The general syntax for multiple inheritance is:

class A:
    Class_Suit of A
    :::
class B:
    Class_Suit of B
    :::
class C(A,B)
    Class_Suit of C
    :::

In multiple inheritance notice that:

- *More than one parent class* for a single child Class.
- The class to be inherited need to be defined first.
- This class is defined like any normal class.
- The definition of the class that will inherit is the same as any normal class. except both the super classes are passed as a parameter.
- The child class can access all the properties of both the parent classes.

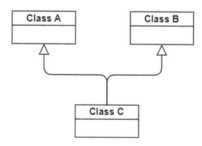

FIGURE 11.4 Multiple inheritance

Assume that the class *GeneralManager* must understand everything from *FinanceManager* as well from *ProductionManager*. So, the *GeneralManager* class must inherit from *ProductionManager* class and *FinanceManager* class.

Example 11.3 Multiple inheritance

```
class FinManager:
  def FinMgrCanDo(self):
    print("Financial manager can do.....")

class ProdManager:
  def ProdMgrCanDo(self):
    print("Production manager can do.....")

class GeneralManager(FinManager,ProdManager):
  def GenMgrCanDo(self):
    print("General manager can do...")

print("\n....Financial Manager Object Created....")
FinM1 = FinManager()
FinM1.FinMgrCanDo()

print("\n\n....Production Manager Object Created....")
ProdM1 = ProdManager()
ProdM1.ProdMgrCanDo()

print("\n...General Manager Object Created...")
GM1 = GeneralManager()
GM1.GenMgrCanDo()
GM1.ProdMgrCanDo()
GM1.FinMgrCanDo()
```

Observe that the object GM1 of class *GeneralManager* can access all the methods of *ProdManager* as well as *FinManager*.

Output

```
...Financial Manager Object Created...
Financial manager can do...

...Production Manager Object Created...
Production manager can do...

...General Manager Object Created...
General manager can do...
Production manager can do...
Financial manager can do...
```

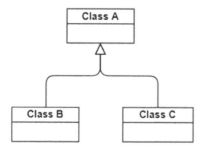

FIGURE 11.5   Hierarchical inheritance

11.1.4   HIERARCHICAL INHERITANCE

As shown in figure 11.5, two different classes may inherit from a single class. In other words, two child classes having a single parent class. This type of inheritance is called *hierarchical inheritance.*

The general syntax for hierarchical inheritance is:

class A:
   Class_Suit of A
   :::
class B(A):
   Class_Suit of B
   :::
class C(A)
   Class_Suit of C
   :::

In a multiple inheritance notice that:

- *More than one child class* for a single parent class.
- The class to be inherited need to be defined first.
- This class is defined like any normal class.
- The definition of the class that will inherit is the same as any normal class except the super class is passed as a parameter.
- The child class can access all the properties of the parent class.

Assume that there is a *Vehicle* class, *Bike* class and *Car* class. The *Bike* is a Vehicle; the *Car* is also a Vehicle. So, the *Bike* and *Car* class inherits the properties of the *Vehicle* Class.

Example 11.4 Hierarchical inheritance

```
class Vehicle:
  def VehicleFunc(self):
    print("Vehicle Properties...")
```

```
class Bike(Vehicle):
  def BikeFunc(self):
    print("Bike Properties...")

class Car(Vehicle):
  def CarFunc(self):
    print("Car Properties...")

print("\n...Vehicle Object Created...")
V1 = Vehicle()
V1.VehicleFunc()

print("\n...Bike Object Created...")
B1 = Bike()
B1.BikeFunc()
B1.VehicleFunc()

print("\n...Car Object Created...")
C1 = Car()
C1.CarFunc()
C1.VehicleFunc()
```

Output

```
...Vehicle Object Created...
Vehicle Properties...

...Bike Object Created...
Bike Properties...
Vehicle Properties...

...Car Object Created...
Car Properties...
Vehicle Properties...
```

11.1.5   HYBRID INHERITANCE

It may happen that we can form an inheritance by combining any of the above four types of inheritance. This type of inheritance is called *hybrid inheritance.* In other words, it is a mixed inheritance. How the mixing of inheritance can be done is left to readers' imagination.

11.2   INITIALIZATION IN INHERITANCE

We have already learnt about the initialization of an attribute in an object in the previous chapter. To refresh, a constructor initializes an object while creating it. We need not call the initialization method (*__init__()*) explicitly. We understand that while talking about inheritance there is an existence of more than one class. Based on which class the constructor is defined, parent or child class of single or multi-level inheritance, there can be various possibilities.

1. __*init*__() for child class only is defined
2. __*init*__() for the parent class only is defined.
3. __*init*__() for both child and parent is defined.

11.2.1 __ INIT __ () FOR THE CHILD CLASS ONLY IS DEFINED

When the __*init*__() method for only the child class (sub class) is defined, it automatically calls this method of the child class.

Example 11.5

```
class Employee():
  def EmpCanDo(self):
    print("Employee can do...")

class Manager(Employee):
  def __init__(self):
    print("Object of Manager Created")
  def MgrCanDo(self):
    print("Manager can do...")

print("\n...Inside Employee Object...")
Emp1 = Employee()

print("\n...Inside Manager Object...")
Mgr1 = Manager()
```

Observe that here, *Employee* is the parent class and *Manager* is the child class. The __*init*__() is defined only for the child class. This method is called while creating the object of manager.

Output:

```
...Inside Employee Object...

...Inside Manager Object...
Object of Manager Created
```

11.2.2 __ INIT __ () FOR THE BASE CLASS ONLY IS DEFINED

When the __*init*__() method for only the parent class (base class) is defined, it automatically calls this method of the base class.

Example 11.6

```
class Employee():
  def EmpCanDO(self):
    print("Employee can do...")
```

```
class Manager(Employee):
  def __init__(self):
    print("Object of Manager Created")
  def MgrCanDo(self):
    print("Manager can do...")

print("\n...Inside Employee Object...")
Emp1 = Employee()

print("\n...Inside Manager Object...")
Mgr1 = Manager()
```

Observe that we are having a __init__() method defined inside the *Employee* class (parent class). While the object is created for *Employee* (parent class) and *Manager* (child class) class the __init__() of the parent class is called automatically.

Output

```
...Inside Employee Object...
Object of Employee Created

...Inside Manager Object...
Object of Employee Created
```

11.2.3   _ _ INIT _ _ () FOR BOTH SUB CLASS AND SUPER CLASS IS DEFINED

When the __init__() method for both child and parent is defined while creating an object of any class, it automatically calls the respective method. However, while calling this method of the parent class, we need to do so explicitly by using the keyword *super()*.

Example 11.7 This example shows if the __init__() method is defined in both parent and child class, while creating the object, only the respective constructor is called.

```
class Employee():
  def __init__(self):
    print("Object of Employee Created")
  def EmpCanDo(self):
    print("Employee can do....")

class Manager(Employee):
  def __init__(self):
    print("Object of Manager Created")
  def MgrCanDo(self):
    print("Manager can do...")

print("\n ...Inside Employee Object...")
Emp1 = Employee()
```

```
print("\n...Inside Manager Object...")
Mgr1 = Manager()
```

Observe that here the constructor __init__() is defined for the child class (Manager) as well as the parent class (Employee). The respective constructor is called while creating the object.

Output

```
...Inside Employee Object...
Object of Employee Created

...Inside Manager Object...
Object of Manager Created
```

Example 11.8 This example shows the use of keyword *super()* while calling the constructor of the parent class.

```
class Employee():
  def __init__(self):
    print("Object of Employee Created")
  def EmpCanDo(self):
    print("Employee can do....")

class Manager(Employee):
  def __init__(self):
    super().__init__()
    print("Object of Manager Created")
  def MgrCanDo(self):
    print("Manager can do....")

print("\n...Inside Employee Object....")
Emp1 = Employee()

print("\n....Inside Manager Object....")
Mgr1 = Manager()
```

Observe that here the __init__() method is defined for the child class (Manager) as well as the parent class (Employee). It is clear that while creating the object the __init__() method of the respective class is called. But if we want to call this method defined in the parent class, we need to use a keyword *super()* inside the __init__() method of the child class.

Output

```
...Inside Employee Object...
Object of Employee Created

...Inside Manager Object...
Object of Employee Created
Object of Manager Created
```

While using _ _ init _ _ () method in multiple inheritance, if the constructors are defined in all the classes (parent class as well as child class), the constructor of the respective class will be called.

Example 11.9

```
class FinManager():
  def __init__(init):
    print("init in the object of Finance Manager")

class ProdManager():
  def __init__(init):
    print("init in the object of Production Manager")

class GeneralManager(FinManager, ProdManager):
  def __init__(init):
    print("init in the object of General Manager")

print("\n...Object of Finance Manager Created...")
Fin1 = FinManager()

print("\n...Object of Production Manager Created...")
Prod1 = ProdManager()

print("\n...Object of General Manager Created...")
Gen1 = GeneralManager()
```

Observe that in the parent class (FinManager and ProcManager) constructors are defined. In the child class (GeneralManager) also a constructor is defined. While creating the objects, their respective constructors get called.

Output

```
...Object of Finance Manager Created...
init in the object of Finance Manager

...Object of Production Manager Created...
init in the object of Production Manager

...Object of General Manager Created...
init in the object of General Manager
```

But if we are using the *super()* keyword in the child class in order to access the constructor of the parent class, it can be called based on the concept called *Method Resolution Order* (MRO). Details on MRO are discussed in the next section.

Example 11.10

```
class FinManager():
  def __init__(init):
    print("init in the object of Finance Manager")
```

```
class ProdManager():
  def __init__(init):
    print("init in the object of Production Manager")

class GeneralManager(FinManager, ProdManager):
  def __init__(init):
    super().__init__()
    print("init in the object of General Manager")

print("\n...Object of Finance Manager Created...")
Fin1 = FinManager()

print("\n...Object of Production Manager Created...")
Prod1 = ProdManager()

print("\n...Object of General Manager Created...")
Gen1 = GeneralManager()
```

Observe that it is a multiple inheritance in which the child class *GeneralManager* is created by inheriting from the parent classes *FinManager* and *ProdManager*. Here the order is important. Observe the parameters in the statement *GeneralManager(FinManager, ProdManager)*. The first parameter is *FinManager* and the *__init__()* method of this class only is called when the keyword *super()* is used.

Output

```
...Object of Finance Manager Created...
init in the object of Finance Manager

...Object of Production Manager Created...
init in the object of Production Manager

...Object of General Manager Created...
init in the object of Finance Manager
init in the object of General Manager
```

11.3   METHOD RESOLUTION ORDER

Method resolution order defines the order in which the base classes are searched while executing a method. First, the method or attribute is searched within a class and then it follows the order we specified while inheriting. This order is also called *linearization of class* and the set of rules are called *method resolution order*. While inheriting from another class, the interpreter needs a way to resolve the methods that are being called via an instance. It plays a vital role in the context of multiple inheritances as the method with the same name may be found in multiple parent classes.

Method resolution order characterizes the request where the base classes are looked through when executing a method. Initially, the methods are looked inside a class and then it follows the order we indicated while inheriting.

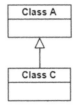

FIGURE 11.6 Method resolution order in single inheritance

For any class the MRO can be found by the built-in method *mro()* or *__mro__*. The simple rule Python applies to resolve MRO (known as a good head question) says, *in MRO we access each super class level-wise from left-to-right.* It means first the method of the child class, then its parent's methods and then its grandparent's methods can be accessed. There can be several cases while resolving the order in which the methods are called starting with the base class.

Case 1 Single inheritance

If a class has only one parent class, it is known as *single inheritance*. According to *mro()* first the methods of the child class are accessed and then the methods of the parent can be accessed. It is shown in figure 11.6.

Example 11.11

```
class A():
  pass
class C(A):
  pass

print(C.mro())
```

Observe that here the child class is *Class C* and the parent class is *Class A*. According to *mro()* first the method of the child class that is *Class C* is called and then its parent class, that is *Class A*, is called.

Output

```
[<class '__main__.C'>, <class '__main__.A'>, <class 'object'>]
```

Case 2 Multi-level inheritance

In this case, there are at least three classes i.e., child, parent and grandparent. At level 0, we have the child class, at level 1 we have parent class and at level 2, we have grandparent class. We can access the methods in the same order.

Example 11.12

```
class A():
  pass
```

```
class B(A):
  pass

class C(B):
  pass

print(C.mro())
```

Here *Class C* is the child class which is at level 0, which inherits from *Class B* which is the parent class at level 1. And *Class B* inherits from *Class A* which is the grandparent class at level 2. Observe that the order in which we resolve the methods, we first access the class at level 0, next the methods at level 1 and last the method at level 2.

Output

```
[<class '__main__.C'>, <class '__main__.B'>, <class '__
main__.A'>, <class 'object'>]
```

Case 3 Multiple inheritance

In this case, the child class has at least two parents. The child class is at level 0 and the parents at level 1. According to MRO first we can access the methods of a class at level 0 and then the methods of a class at level 1. As more than one class at level 1, now it will be resolved from left-to-right.

Example 11.13

```
class A():
  pass

class B():
  pass

class C(A,B):
  pass

print(C.mro())
```

Observe that *Class C* has two parent classes *Class A* and *Class B*. According to MRO first the methods for the child class (Class C) at level 0 are accessed and the methods of parent class at level 1 are accessed from left-to-right. The concept of left-to-right comes from the order of calling the class while inheriting. Observe the statement *class C(A, B)*. First, class A is called and then class B is called. In this order only the methods can be called.

Output

```
[<class '__main__.C'>, <class '__main__.A'>, <class '__
main__.B'>, <class 'object'>]
```

Case 4(i): Multiple and multi-level inheritance (mixed inheritance)

To find the order in which methods are accessed in a mixed inheritance, the rule is the same, bottom-to-top and left-to-right (figure 11.7).

Example 11.14

```
class A():
  pass

class B():
  pass

class C(A,B):
  pass

class D(C,B):
  pass

print(D.mro())
```

Observe that here the child class is *Class D* which is at level 0. Its parent class is *Class C* and *Class B* which are at level 1. And at last, we have *Class A* which is at level 2. Accordingly, the respective methods are called. In this case there are two classes at level 1, it is accessed from left-to-right.

Output

```
[<class '__main__.D'>, <class '__main__.C'>, <class '__
main__.A'>, <class '__main__.B'>, <class 'object'>]
```

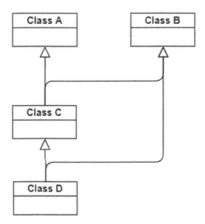

FIGURE 11.7   Method resolution order in mixed inheritance

Case 4(ii): Multiple and multi-level inheritance (mixed inheritance)

In this variant of mixed inheritance also the rule remains the same, bottom-to-top and left-to-right. This type of inheritance is known as *diamond inheritance* (figure 11.8).

Example 11.15

```
class A():
  pass

class B(A):
  pass

class C(A):
  pass

class D(B,C):
  pass

print(D.mro())
```

Observe that *Class D* is at level 0, which is the child class. It has two parents *Class B* and *Class C* at level 1. These two classes at level 1 have a single parent *Class A* at level 2. The order of resolving access of methods is first *Class D*, then its parents at level 1 from left-to-right that is *Class B* and *Class C*. At last *Class-A* at level 2.

Output

```
[<class '__main__.D'>, <class '__main__.B'>, <class '__main__.C'>, <class '__main__.A'>, <class 'object'>]
```

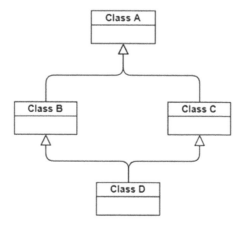

FIGURE 11.8   Method resolution order in diamond inheritance

11.4   SPECIALIZED METHODS

There are two specialized built-in methods available to deal with inheritance. The methods are discussed here.

11.4.1   ISSUBCLASS()

This method returns True, if a class is sub class of a specific class otherwise it returns False. The general syntax is: ***issubclass(SubClass_name, Class_Name)***.

It requires two parameters: the name of the sub class and the name of the class.

Example 11.16

```
class A:
  pass
class B(A):
  pass
class C:
  pass
x = issubclass(B,A)
print(x)
y = issubclass(C,A)
print(y)
```

Here *Class B* is inherited from *Class A*. Observe that *issubclass(B, A)* returned *True* and *issubclass (C,A)* returned *False*.

Output

```
True
False
```

11.4.2   _ _ BASES _ _

This attribute returns the parent of a child class. If the class does not have any parent class it will return the class object. The general syntax is: ***Class_Name.__bases__***.

Example 11.17

```
class A:
  pass

class B(A):
  pass

print(A.__bases__)
print(B.__bases__)
```

Observe the output, as class A does not have any parent class, so the output is '(<class 'object'>,) '. Class A is the parent of class B, so it prints '(<class '__main__.A'>,)'.

Output:

```
(<class 'object'>,)
(<class '__main__.A'>,)
```

WORKED OUT EXAMPLES

Example 11.18 Write a Python program which create two classes employee and manager. The manager class inherits from the employee class. The employee has three attributes id, name and pay, and the manager class has four attributes id, name, pay and department. Create a method that calculates pay hikes for the employee by 5% and the pay of manager by 10%. Then print the name and pay of the employee before and after the pay hike and for the manager print the name, pay and department before and after the pay hike.

```
class Employee():
  hike = 1.05
  def __init__(self,id,name,pay):
    self.id = id
    self.name = name
    self.pay = pay

  def pay_hike(self):
    self.pay = Float(self.pay * self.hike)

class Manager(Employee):
  hike = 1.10
  def __init__(self,id,name,pay,dept):
    super().__init__(id,name,pay)
    self.dept = dept

Emp1 = Employee(501, "Srusti", 50000)
Emp2 = Manager(502, "Nakshatra",60000, "Finance")
print(Emp1.name,"'s pay ",Emp1.pay)
Emp1.pay_hike()
print(Emp1.name,"'s pay aftr hike ",Emp1.pay)
print("-------\n")
print(Emp2.name,"'s pay ", Emp2.pay," working in ",
Emp2.dept)
Emp2.pay_hike()
print(Emp2.name,"'s pay after hike ",Emp2.pay)
```

Output

```
Srusti 's pay 50000
Srusti 's pay aftr hike 52500.0
-------
Nakshatra 's pay 60000 working in Finance
Nakshatra 's pay after hike 66000.0
```

MULTIPLE CHOICE QUESTIONS

1. Which relation is best suitable for employee and manager?
   a. Association
   b. Inheritance
   c. Composition
   d. None of the above

2. Which relation is best suitable for mango to fruit?
   a. Association
   b. Inheritance
   c. Composition
   d. None of the above

3. Which of the following checks if Class B is inherited for Class A?
   a. B.issubclass(A)
   b. A.issubclass(B)
   c. issubclass(A,B)
   d. issubclass(B,A)

4. Which of the following is true about super()?
   a. Finds the super set
   b. Finds the super class
   c. Access methods of super class
   d. Access methods of sub class

5. Which of the following is not true about __bases__?
   a. Finds the derived class
   b. Finds the super class
   c. Finds the parent class
   d. Finds the of sub class

6. Find the output of the following code
   ```
   class A:
       def __init__(self):
           self.a = 0

   class B(A):
       def __init__(self):
           A.__init__(self)
           self.b = 1
   ```

```
def main():
b1 = B()
print(b1.a,b1.b)

main()
```

  a. 0 1
  b. 1 0
  c. NameError
  d. TypeError

7. Which of the following is not a type of inheritance?
  a. Single-level Inheritance
  b. Double-level Inheritance
  c. Hybrid Inheritance
  d. Multi-level Inheritance

8. Find the output of the following code

```
class A:
    def one(self):
      return self.two()

    def two(self):
      return 'A'

class B(A):
    def two(self):
      return 'B'

obj1 = A()
obj2 = B()
print(obj1.two(),obj2.two())
```

  a. A B
  b. B A
  c. NameError
  d. TypeError

9. Find the type of inheritance

```
class A:
  pass
class B:
  pass
classC(A,B):
  pass
```

  a. Single-level
  b. Multi-level
  c. Multiple
  d. Hybrid

10. Find the type of inheritance
    class A:
      pass
    class B(A):
      pass
    class C(B):
      pass

    a. Single-level
    b. Multi-level
    c. Multiple
    d. Hybrid

11. Single-level inheritance refers to
    a. A sub class that has a single super class
    b. A sub class that has multiple super classes
    c. A super class that has many sub classes
    d. None of the above

12. Find the order in which the methods with same name may be accessed in
    class A:
      pass
    class B():
      pass
    class C(A,B):
      pass
    print(C.mro())

    a. A B C
    b. C B A
    c. A C B
    d. C A B

13. Find the order in which the methods with same name may be accessed in
    class A():
      pass
    class B():
      pass
    class C(A,B):
      pass
    class D(C,B):
      pass
    print(D.mro())

    a. D C A B
    b. D C B A
    c. D A C B
    d. D A B C

14. Which of the following can be used to invoke the __init__ method in B from A, where A is a subclass of B?
    a. super().__init__(self)
    b. B.__init__()
    c. super().__init__()
    d. B.__init__(self)

15. Find the output of the following code
    ```
    class A():
        def show(self):
            print("A show()")

    class B(A):
        pass

    obj = B()
    obj.show()
    ```
    a. Nothing is printed
    b. Prints "A show()"
    c. Error
    d. None of the above

16. Find the output of the following code
    ```
    class A():
        def __init__(self,a=3):
            self._a = a
    class B(A):
        def __init__(self):
            super().__init__(30)
        def display(self):
            print(self._a)

    obj = B()
    obj.display()
    ```
    a. 3
    b. 30
    c. Error
    d. None of the above

17. Find the output of the following code
    ```
    class A:
        def test(self):
            print("Class of A is called")
    class B(A):
        def test(self):
    ```

```
        print("Class of B is called")
    class C(A):
        def test(self):
            print("Class of C is called")
    class D(B,C):
        def test(self):
            print("Class of D is called")

    obj = D()
    obj.test()
```

a. Class of A called
b. Class of B called
c. Class of C called
d. Class of D called

18. Find the output of the following code

```
    class A:
        def test(self):
            print("Class of A is called")

    class B(A):
        def test(self):
            print("Class of B is called")

    class C(A):
        def test(self):
            print("Class of C is called")

    class D(A,C):
        def test(self):
            print("Class of D is called")

    obj = D()
    obj.test()
```

a. Class of A called
b. Class of B called
c. Class of C called
d. TypeError

19. Find the output of the following code

```
    class A:
        def test(self):
            print("Class of A is called")

    class B(A):
        def test2(self):
```

```
        print("Class of B is called")
    class C(A):
        def test2(self):
            print("Class of C is called")

    class D(B,C):
        def test2(self):
            print("Class of D is called")

    obj = D()
    obj.test()
```

a. Class of A called
b. Class of B called
c. Class of C called
d. Class of D called

20. Find the output of the following code

```
    class A:
        def test(self):
            print("Class of A is called")
    class B(A):
        def test1(self):
            print("Class of B is called")

    class C(A):
        def test1(self):
            print("Class of C is called")
    class D(B,C):
        def test2(self):
            print("Class of D is called")

    obj = D()
    obj.test()
```

a. Class of A called
b. Class of B called
c. Class of C called
d. Class of D called

DESCRIPTIVE QUESTIONS

1. What is inheritance? What are the different types of inheritance? Explain with examples.
2. Explain with an example how the constructor of a parent class can be accessed for the child class.
3. What is the use of super()?
4. What is MRO? Explain different types of MRO with example.
5. What is meant by depth-first, left-to-right in multiple inheritance?

PROGRAMMING QUESTIONS

1. Write a program that defines a shape class with a constructor that gives value to width and height. Then define two sub classes, triangle and rectangle, that calculate the area of the shape area (). In the main, define two variables, a triangle and a rectangle and then call the area() function in these two variables.

2. Write a program with a mother class and an inherited daughter class. Both of them should have a method show () that prints a message (different for mother and daughter). In the main define a daughter and call the show() method on it.

3. Write a program with a mother class animal. Inside it, define a name and an age variable and initialize them. Then create two base variables zebra and dolphin which write a message telling their age, name and giving some extra information (e.g., place of origin).

4. Write a Python program that demonstrates how to access the local variables, child class variables, parent class variables and global variables.

5. Write a Python program and create two classes employee and manager. The manager class inherits from the employee class. The employee has two attributes, id and name and the manager class has three attributes, id, name and list of employees supervised by the manager. Implement addition and deletion of employees supervised by the manager.

ANSWER TO MULTIPLE CHOICE QUESTIONS

1.	B	2.	B	3.	D	4.	C	5.	D
6.	A	7.	B	8.	A	9.	C	10.	B
11.	A	12.	D	13.	A	14.	C,D	15.	B
16.	B	17.	D	18.	B	19.	A	20.	A

12 Polymorphism

LEARNING OBJECTIVES

After studying this chapter, the reader will be able to:

- Understand polymorphism
- Know the different implementations in polymorphisms
- Use operator overloading
- Use method overloading
- Apply encapsulation

12.1 INTRODUCTION

Polymorphism is one of many important concepts in object-oriented programming. It tells us how to access and manipulate different attributes and methods without regard to their specific class. Depending on the type of object, the methods act differently. The ability of the method to behave differently depending on the class type is known as *Polymorphism.*

Example 12.1

```
i = 1
print('Datatype of i',type(i),'Value of i is',i)
i=i+1
print('Datatype of i',type(i),'Value of i+1 is',i)
j= True
print('Datatype of j',type(j),'Value of j is',j)
j=j+1
print('Datatype of j',type(j),'Value of j+1 is',j)
```

In the above example observe that *i* is an Integer type data with value 1 and *j* is a Boolean type data with value True. When we add 1 to *i*, it returns 2. At the same time when 1 is added to *j*, it also returns 2. Both *i+1* and *j+1* returned an Integer. This is known as *dynamic binding*. At the time of execution, while adding a Boolean variable to an Integer value, there is a dynamic type conversion from Boolean to Integer and hence the result is an Integer.

Output

```
Datatype of i <class 'int'> Value of i is 1
Datatype of i <class 'int'> Value of i+1 is 2
Datatype of j <class 'bool'> Value of j is True
Datatype of j <class 'int'> Value of j+1 is 2
```

DOI: 10.1201/9781003219125-14

Example 12.2

```
a = "Python"
b = "Learning"
print ('Value of a+b is', a+b)
print ('Type of a+b is', type (a+b))
```

In this example we can observe that there are two variables *a* and *b* with values *"Python"* and *"Learning"* respectively. When we perform addition, we get a new String *PythonLearning*. In the previous example (example- 2.1) we have seen that the + operation adds two numbers. Now we can see that the same + operation concatenates two strings.

Output

```
Value of a+b is PythonLearning
Type of a+b is <class 'str'>
```

Example 12.3

```
i = 3
j = i*3
print (j)
k = '3'
print (k*3)
```

In this example observe that the multiplication (*) operation, when applied to Integers acts like an arithmetic multiplication operation whereas when applied to a String, it acts like a repetition of String.

Output

```
9
333
```

We can observe from the above example that the attributes and methods change their behavior depending on the situation. While inheritance is related to classes and their hierarchy, polymorphism on the other hand, is related to methods. When polymorphism is applied to a method, depending on the given parameters, a particular form of the function can be selected for execution. In Python, method overriding is one way of implementing polymorphism.

There are four different implementations in polymorphisms:

1. Duck-typing
2. Operator overloading
3. Method overloading
4. Method overriding

12.1.1 DUCK-TYPING

We have seen in the previous examples that in Python we cannot specify the type explicitly. Based on the data provided at runtime, the type is decided automatically. Hence, Python is considered a Dynamically Typed Programming Language.

Python follows a concept called *Duck Typing Philosophy*. According to this, if a bird walks like a duck and talks like a duck, it must be a duck. In duck typing we do not care if this object is actually a duck or not, we only care if it behaves like a duck. We simply don't care what type of object we're working with; we only care if our object can perform a specific operation based on the datatype.

Example 12.4

```python
class HomeMaker:
  def worksAt(self):
    print("Works at Home")
  def worksFor(self):
    print("For the Family")

class Employee:
  def worksAt(self):
    print("Works at Office")
  def worksFor(self):
    print("For the Employer")

class Farmer:
  def worksAt(self):
    print("Works in the Field")
  def worksFor(self):
    print("For the Nation")

def workOf(person):
  person.worksAt()
  person.worksFor()

i = [HomeMaker(), Employee(), Farmer()]
for person in i:
  workOf(person)
  print("------")
```

Observe that there are three classes defined namely, *HomeMaker*, *Employee* and *Farmer*. In each class two methods are defined with the same name, *worksAt()* and *workFor()*. There is a method *workOf(person)* defined in which *worksAt()* and *workFor()* are called. Both functions are activated regardless of the object which is passed to *workOf()* method.

Output

```
Works at Home
For the Family
------
Works at Office
For the Employer
------
Works in the Field
For the Nation
------
```

It may happen that the method we are trying to invoke is not present in some class. In that case, the interpreter may raise and attribute an error.

Example 12.5

```
class HomeMaker:
  def worksAt(self):
    print("Works at Home")

class Employee:
  def worksAt(self):
    print("Works at Office")

class Farmer:
  def worksFor(self):
    print("For the Nation")

  def workOf(person):
  person.worksAt()

i = [HomeMaker(), Employee(), Farmer()]
for person in i:
  workOf(person)
  print("_____")
```

Observe that in the class *HomeMaker()* and *Employee(),* we have a method *worksAt(),* whereas in the *Farmer()* class, we don't have *workAt().* When we execute the code, it throws an *AttributeError.*

Output

```
Works at Home
_____
Works at Office
_____
Traceback (most recent call last):
```

```
File "C:\Users\RakeshN\AppData\Local\Programs\Python\Py
thon37-32\chap-12.py", line 96, in <module>
  workOf(person)
File "C:\Users\RakeshN\AppData\Local\Programs\Python\Py
thon37-32\chap-12.py", line 92, in workOf
  person.worksAt()
AttributeError: 'Farmer' object has no attribute 'worksAt'
```

In order to avoid such an *AttributeError*, we can take permission before executing the code. The permission is nothing but a check in which we check if the attribute is existing in a class. If the attribute exists, then only we can execute it. Such a way of taking permission for executing a code is known as *Non-Pythonic*.

Example 12.6

```
class HomeMaker:
  def worksAt(self):
    print("Works at Home")

class Employee:
  def worksAt(self):
    print("Works at Office")

class Farmer:
  def worksFor(self):
    print("For the Nation")

  def workOf(person):
    person.worksAt()

i = [HomeMaker(), Employee(), Farmer()]
for person in i:
  if hasattr(person,"worksAt"):
    workOf(person)
    print("_____")
```

Observe that in the code, we place a conditional statement *hasattr(person, 'workAt')*. It will first check for which objects this condition is true. Only for those objects, *workOf()* method is executed. In other words, we are taking permission before executing the code for an object. This will not throw any error but we are unaware of the reason why the method *worksAt()* did not work.

Output

```
Works at Home
_____
Works at Office
_____
```

Instead, we can use a way which is *Pythonic*. It is easier to ask for forgiveness than permission. It means we just try to do something and if it doesn't work then we'll handle it.

Example 12.7

```
class HomeMaker:
  def worksAt(self):
    print("Works at Home")

class Employee:
  def worksAt(self):
    print("Works at Office")

class Farmer:
  def worksFor(self):
    print("For the Nation")

def workOf(person):
  try:
    person.worksAt()
  except AttributeError as err:
    print(err)

i = [HomeMaker(), Employee(), Farmer()]
for person in i:
  workOf(person)
  print("--------")
```

Observe the *workOf()* method. Irrespective of the type of object, we call the method *worksAt()*. The error is handled by *try... except* construct. For the time being, we don't bother with it, we are going to discuss it in detail in Chapter 14. Now we know the reason why the *worksAt()* did not activate. That is the reason we say it's easier to ask for forgiveness than permission. It means, try to do something and if it works then great and if not then just handle that error.

Output

```
Works at Home
--------
Works at Office
--------
'Farmer' object has no attribute 'worksAt'
--------
```

12.1.2 OPERATOR OVERLOADING

In example 12.1–12.3, we have seen that the addition (+) operation can be used to add two numbers and returns the sum of two numbers. The same addition (+)

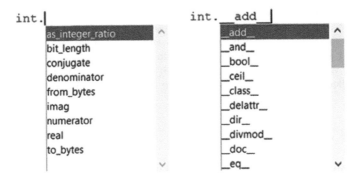

FIGURE 12.1   Operator overloading

operation can also be used to add two strings and return the concatenation of the two strings.

Even though these actions can be implemented via class methods, using overloading is closer to Python's object model and the object interfaces are more consistent with Python's built-in objects. Hence, overloading is easier to learn and use. For every operator *magic methods* are available. To overload any operator, we have to override that method in a class.

For example, we can get the set of methods that are applicable to Integers and that can be overloaded. We can do so by writing *int.* (int followed by a dot), we get a pop-up drop-down list with all the methods that can be overloaded. (otherwise, to get this drop-down list press *ctrl+space* after the dot). This is shown in figure 12.1.

Example 12.8

User-defined classes can override nearly all of Python's built-in methods. These methods are identified by having two underscores before and after the method name, such as __add__. These methods are automatically called when Python performs on operators; when an expression has "+" in it, the user's method will be used instead of Python's built-in method __add__.

Example 12.9

```
class Employee():
  def __init__(self, id, sal):
    self.Eid = id
    self.salary = sal

  def __add__(self, other):
    return self.salary + other.salary
Emp1 = Employee(501, 4000)
Emp2 = Employee(502, 7000)

print("Sum of salary is ", Emp1 + Emp2)
```

Here we have a class with two attributes namely *Eid* and *salary*. To find the sum of salaries, we printed *Emp1 + Emp2*. But these two are objects and we don't have any method that adds objects. We will get an error message *+ is not supported between instances of Employee*. In order to get the sum of an attribute in the objects, we define a method *__add__* with objects as arguments. It means we have called *Employee.__add__(s1,s2)* where *s1* is self and *s2* is other.

Output

```
Sum of salary is 11000
```

Similarly, we can define many such functions for overloading. Before performing the overloading, we need to check if we are having a function which starts and end with two underscores.

12.1.3   METHOD OVERLOADING

If we call a method with different parameters, we call it *method overloading*. We know that every method has either zero or a few arguments. In order to call the method, every time we need to pass the same number of parameters. If we are allowed to pass the same or a smaller number of parameters (not fixed) than defined and still we can execute the same method, we call the method *overloaded*.

```
class Class_name():



def Method(self, a, b, c, d):
    :
    :
    :
    :
Res = Class_name.Method(a)
Res = Class_name.Method(a,b)
Res = Class_name.Method(a,b,c)
Res = Class_name.Method(a,b,c,d)
```

Observe that here we have the method *Class_name.Method()* which takes four parameters. While calling the method we can pass one, two, three or four parameters i.e., the method is called with different parameters. This is shown in figure 12.2.

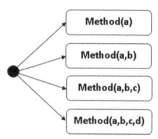

FIGURE 12.2   Method overloading

Unlike other object-oriented programming languages, Python does not support direct method overloading. We need to make some adjustments to achieve method overloading in Python.

Example 12.10

```
class Employee():
  def Salary(self, basic = None, hra = None, da= None,
  housing = None):
  salary = 0
  if (basic != None and hra != None and da != None and
  housing != None):
    salary = basic + hra + da + housing
  elif(basic != None and hra != None and da != None):
    salary = basic + hra + da
  elif(basic != None and hra != None):
    salary = basic + hra
  else:
    salary = basic
  return salary

Emp1 = Employee()
sal = Emp1.Salary(12000, 1200, 12000, 800)
print("Salary of the Employee (basic + hra + da + housing)
is:", sal)

sal = Emp1.Salary(12000, 1200, 12000)
print("Salary of the Employee (basic + hra + da) is: ", sal)

sal = Emp1.Salary(12000,1200)
print("Salary of the Employee (basic + hra) is: ", sal)

sal = Emp1.Salary(12000)
print("Salary of the Employee (basic) is: ", sal)
```

Observe that the arguments for the method *Salary()* are (*self, basic, hra, da, housing*). Out of these arguments, *self* is the default argument and for the rest of the arguments we have default values for the parameters: *basic = None, hra = None, da = None, housing = None*. We need to include *if...elif...else* constructs in order to handle the situation of not passing value of any parameter.

Output

```
Salary of the Employee (basic + hra + da + housing) is:
26000
Salary of the Employee (basic+hra+da) is: 25200
Salary of the Employee (basic + hra) is: 13200
Salary of the Employee (basic) is: 12000
```

However, we can implement method overloading slightly differently by passing the argument list. To deal with a list of arguments we have *args*. It is a special syntax definition in Python which is used to pass a variable number of arguments to a method. It is used to pass a *non-keyword*, *variable-length* argument list. The *args* allows us to take in any number of extra arguments (including zero arguments) than the number of formal arguments that we defined.

Example 12.11

```
class Employee():
  def Salary(self, *args):
    salary = 0
  for x in args:
    salary = salary + x
  return salary

Emp1 = Employee()
sal = Emp1.Salary(12000, 1200, 12000, 800)
print("Salary of the Employee (basic + hra + da + housing)
is: ", sal)

sal = Emp1.Salary(12000, 1200, 12000)
print("Salary of the Employee (basic + hra + da) is: ", sal)

sal = Emp1.Salary(12000, 1200)
print("Salary of the Employee (basic + hra) is: ", sal)

sal = Emp1.Salary(12000)
print("Salary of the Employee (basic) is: ", sal)
```

Observe the method, apart from self (the default parameter) we have one more argument *agrs*. The problem here is readability. We don't know for which variable what value is passed.

Output

```
Salary of the Employee (basic + hra + da + housing) is:
26000
Salary of the Employee (basic + hra + da) is: 25200
Salary of the Employee (basic + hra) is: 13200
Salary of the Employee (basic) is: 12000
```

12.1.4 METHOD OVERRIDING

Method overriding is one of the most important feature in object-oriented programming. Method overriding is basically used in inheritance. In inheritance we have already seen that whichever member is available in parent class, is also available to the child class by default. We can call a method from a child class which is defined in the parent class. For example, if we don't have a phone, we use our parent's phone if we want to talk to someone. If we have our own phone, we never use our parent's phone. If the child class also has the same method (same method name with same parameters) as the method defined in the parent, then we call this *method override*. If we want to call this method from child class, which is overridden, then the method defined in the child class only will be available. *According to MRO*, as discussed in 11.3, the method defined in the child class only will be called. Method overriding is used when we want to modify the existing behavior of the parent class.

Example 12.12

```
class Employee():
  def __init__(self):
    print("Object of Employee Created")
  def EmpCanDo(self):
    print("Employee can do...")

class Manager(Employee):
  def EmpCanDo(self):
    print("Manager can do...")

print("\n...Inside Employee Object...")
Emp1 = Employee()
Emp1.EmpCanDo()

print("\n...Inside Manager Object...")
Mgr1 = Manager()
Mgr1.EmpCanDo()
```

Observe that in the class *Employee()* there are two methods defined *__init__ (self)* and *EmpCanDo(self)* and in the *Manager* class one method defined *EmpCanDo(self)* and the *Manager* class inherits *Employee* class.

In the class *Manager*, as we don't have *__init__()*, so the object of Manager calls *__init__()* method of the *Employee* class because of inheritance. In both classes we have a method *EmpCanDo()* (they may work differently but the method name is the same). When the child class (*Manager*) calls this function, it will override this method in the parent class and execute it in the child class only.

Output

```
...Inside Employee Object...
Object of Employee Created
Employee can do...


...Inside Manager Object...
Object of Employee Created
Manager can do...
```

If we want to call the method which is overridden in the child class, we can use the *super()* keyword.

12.2   ENCAPSULATION

This is an important concept in object-oriented programming languages in which we hide the variables and the actual implementation of a method in which the variables are used. An example of encapsulation is *print()* statement. We know what parameters are required and what output we get to print a statement but we don't know how the *print()* statement does it. It means we can restrict access to the variables and methods so that we can prevent accidental data modification.

Example 12.13

```python
class Mouse():
  def __init__(self,make,type):
    self.make = make
    self.type = type
  def __str__(self):
    return "Mouse :\n\tMade by :"+ self.make +"\n\tType:"+
    self.type

class KeyBoard():
  def __init__(self,make,type):
    self.make = make
    self.type = type

  def __str__(self):
    return "\nKeyboard :\n\tMade by :" + self.make +
    "\n\tType:" + self.type

class Computer:
  def __init__(self, mouse, keyboard):
    self.mouse = mouse
    self.keyboard = keyboard
  def __str__(self):
    return ("\nThe Computer has \n"+str(self.mouse)+" Mouse
    \n"+str(self.keyboard)+" keyboard")

m = Mouse("Logitech", "Wireless")
print(m)
k = KeyBoard("HP", "Wired")
print(k)
comp = Computer("Samsung", "Zebronics")
print(comp)
comp1 = Computer(m,k)
print(comp1)
```

In this program we have three classes *Mouse, KeyBoard* and *Computer*. Each class has its own attributes and methods. We have created the object of Mouse, *m*, and object of KeyBoard, *k*, by passing the make and type as parameters. When the first object of the computer class, *comp*, is created we passed the make of mouse and keyboard. Now the *comp* object behaves like any other object. While creating the second object of the computer class, *comp1*, we passed the object of *Mouse* and *KeyBoard* class.

Observe that when we passed the object as a parameter, the implementation inside their respective class is hidden. Once the variables inside the object of Mouse and the object of KeyBoard are assigned with some values, we cannot change them by calling the object of these classes. We can call a method without knowing the implementation. This is how to hide them and prevent them from accidental data modification.

Output

```
Mouse  :
        Made by :Logitech
        Type:Wireless

Keyboard :
        Made by :HP
        Type:Wired


The Computer has
Samsung Mouse
Zebronics keyboard

The Computer has
Mouse  :
        Made by :Logitech
        Type:Wireless Mouse

Keyboard :
        Made by :HP
        Type:Wired keyboard
```

We can not only call the attributes and methods class defined earlier without using inheritance or method overriding but we can also extend the new implementation. This is also known as *Dynamic Encapsulation.*

Example 12.14

```
class Mouse():
  def __init__(self,make,typ):
    self.make = make
    self.type = typ

  def __str__(self):
    return "Mouse :\n\tMade by :"+ self.make +"\n\tType:"+
    self.type
```

```
class KeyBoard():
  def __init__(self,make,typ):
    self.make = make
    self.type = typ

  def __str__(self):
    return "\nKeyboard :\n\tMade by :" + self.make + "\n\
    tType:" + self.type

class Computer:
  def __init__(self,mouse,keyboard,monitor):
    self.mouse = mouse
    self.keyboard = keyboard
    self.monitor = monitor

  def __str__(self):
    return ("\nThe Computer has \n"+ str(self.mouse)+"
    Mouse\n"+ str(self.keyboard)+" keyboard\n"+"Monitor
    "+self.monitor)

m = Mouse("Logitech", "Wireless")
print(m)
k = KeyBoard("HP", "Wired")
print(k)

comp = Computer("Samsung", "Zebronics", "Acer")
print(comp)
comp1 = Computer(m,k, "SAMSUNG")
print(comp1)
```

The Mouse and KeyBoard class remained the same. In the Computer class now we have to pass three parameters. While creating *comp* object, we passed the makers of mouse, keyboard and monitor. It behaved like any other object. While creating the *comp1* object we have passed the objects of *Mouse* and *KeyBoard*. Apart from that we passed the maker of monitor. Out of three parameters, two are objects and one is local to this object. Observe that we can not only use the methods defined in other class (without knowing their implementation) but also, we can extend any new methods as desired.

Output

```
Mouse :
        Made by :Logitech
        Type:Wireless

Keyboard :
        Made by :HP
        Type:Wired
```

```
The Computer has
Samsung Mouse
Zebronics keyboard
Monitor Acer

The Computer has
Mouse :
        Made by :Logitech
        Type:Wireless Mouse

Keyboard :
        Made by :HP
        Type:Wired keyboard
        Monitor SAMSUNG
```

WORKED OUT EXAMPLES

Example 12.15 Write a Python program that prints the bird, its pet name and its sound using polymorphism

```python
class Bird:
  def __init__(self, name=""):
    self.name = name
    def sound(self):
      pass

class Parrot(Bird):
  def __init__(self,name=""):
    super().__init__(name)
  def sound(self):
    print(self.name," is a Parrot and it can talk")

class Rooster(Bird):
  def __init__(self, name=""):
    super().__init__(name)
  def sound(self):
    print(self.name," is a Rooster it says Cookdu-Coo")

parrotBird= Parrot("Mithoo")
parrotBird.sound()

roosterBird = Rooster("Mithoo")
roosterBird.sound()
```

Output

```
Mithoo is a Parrot and it can talk
Mithoo is a Rooster it says Cookdu-Coo
```

Example 12.16 Write a Python program that overloads a method that prints the sum and two operands or three operands are passed but when one operand is passed prints sum cannot be performed with one operand

```python
class MethOvrLod():
  def sum(self, a=None, b=None, c=None):
    if(a!=None and b!=None and c!=None):
      print("The sum of 3 numbers is : ", a+b+c)

    elif (a!=None and b!=None):
      print("The sum of 2 numbers is :", a+b)

    else:
      print("The sum cannot be performed with single
      operand.")

obj = MethOvrLod()
obj.sum(100,200,300)
obj.sum(20,30)
obj.sum(9)
```

Output

```
The sum of 3 numbers is : 600
The sum of 2 numbers is : 50
The sum cannot be performed with single operand.
```

MULTIPLE CHOICE QUESTIONS

1. Which function overloads + operator?
   a. __add__()
   b. __plus__()
   c. __sum__()
   d. None of the above

2. Which operator is overloaded by __invert__()?
   a. ^
   b. !
   c. ~
   d. None of the above

3. Which function overloads == operator?
   a. __isequal__()
   b. __equal__()
   c. __eq__()
   d. __equ__()

4. Which function overloads < operator?
   a. __lt__()
   b. __less__()
   c. __eq__()
   d. __le__()

5. Which operator is overloaded by __gl__()?
   a. >
   b. >=
   c. <>
   d. None of the above

6. Which function overloads << operation?
   a. __lt__()
   b. __ge__()
   c. __ rshift __()
   d. __lshift__()

7. Which function overloads I operation?
   a. __()
   b. _()
   c. _()_
   d. __()__

8. Which operator is overloaded by __floordiv__()?
   a. /
   b. //
   c. \\
   d. |

9. Can a subclass modify the behavior of its superclass?
   a. Yes
   b. No
   c. Depends on the situation
   d. None of the above

10. Polymorphism refers to
    a. Multiple existences of the same class
    b. Multiple existences of the same object
    c. Multiple behaviors of the same method
    d. None of the above

11. The code below does not raise any error because of
    print(5 + 10)
    print('5' + '10')

    a. Polymorphism
    b. Isomorphism
    c. The compiler overlooks it
    d. None of the above

12. Method overriding is achieved by implementing
    a. Polymorphism
    b. Isomorphism
    c. Inheritance
    d. None of the above

13. Method overriding is possible when
    a. The parent and the child have a method with a different name
    b. The parent and the child have a method with the same name
    c. Machine dependent
    d. None of the above

14. In method overloading
    a. Any number of arguments can be passed
    b. A fixed number of arguments can be passed
    c. Zero arguments canbe passed
    d. None of the above

15. Duck typing refers to
    a. Making the code smaller
    b. More restrictions on the type of values that can be passed to a given
       method
    c. Less restrictions on the type of values that can be passed to a given
       method
    d. No restriction on the type of values that can be passed to a given method

16. Find the output of the following code
    ```
    class A:
        def __repr__(self):
            return "1"
    class B(A):
        def __repr__(self):
            return "2"
    class C(B):
        def __repr__(self):
            return "3"
    obj1 = A()
    obj2 = B()
    obj3 = C()
    print(obj1, obj2, obj3)
    ```

a. 111
b. 222
c. 333
d. 123

17. Find the output of the following code
```
class A:
    def check(self):
        return "Check Inside A"
    def display(self):
        print(self.check())
class B(A):
    def check(self):
        return "Check Inside B"
def main():
    A().display()
    B().display()
main()
```

a. Check Inside A and Check Inside B
b. Check Inside B and Check Inside A
c. Syntax Error
d. ValueError

18. Find the output of the following code
```
class A:
    def Hi(self):
        return self.Hello()
    def Hello(self):
        return 'A'

class B(A):
    def Hello(self):
    return 'B'

obj2 = B()
print(obj2.Hello())
```

a. A
b. B
c. A B
d. B A

19. Find which of the following statement is false
a. All the private methods can be overriden
b. Only a few private methods can be overriden

    c. Only the public methods can be overriden

    d. Only a few public methods can be overridden

20. Find which of the following statement is false

    a. Method overloading takes zero or more arguments

    b. Magic method is those methods, which start and end with __ (double underscore)

    c. Operator overloading is done to magic methods

    d. Typing "a duck" is called duck typing

DESCRIPTIVE QUESTIONS

1. Describe the different ways of implementing polymorphism.
2. What is duck typing? What is its utility?
3. Describe the conditions under which method overloading can be done.
4. What is operator overloading? Explain with an example.
5. What is overriding? Explain with an example

PROGRAMMING QUESTIONS

1. Write a Python program that manages the bank account with two operations defined i.e., deposit() and withdraw(). There are two different types of accounts i.e., current account and savings account with a minimum deposit of 1500 and 1000 respectively. When some amount is deposited to a savings account interest of 5% is gained on the newly deposited amount. When an amount greater than 1000 is withdrawn from the current account a message appears asking for special permission.
2. Write a Python program that overloads a method which print the sum of any number of Integers and prints an appropriate result. When one number is passed it prints a sum that cannot be performed with one operand.
3. Write a Python program that overrides the __str__ method.

ANSWER TO MULTIPLE CHOICE QUESTIONS

1.	A	2.	C	3.	C	4.	A	5.	D
6.	D	7.	A	8.	B	9.	A	10.	C
11.	A	12.	C	13.	B	14.	A	15.	C
16.	D	17.	A	18.	B	19.	C	20.	D

13 Abstract Class, Aggregation, Composition

LEARNING OBJECTIVES

After studying this chapter, the reader will be able to:

- Understand abstract class and abstract method
- Know different types of relationships and their uses
- Differentiate between inheritance and association
- Differentiate between aggregation and composition

13.1 INTRODUCTION – ABSTRACT CLASS

Until now the concepts we have discussed are concrete, a concrete class, a concrete method, etc. A concrete class is a class wherein we can instantiate, and a method is called a concrete method where we implement. Abstraction is a process of hiding details from the user.

An example is a remote control of a television. The remote control provides a set of functionalities and the way to operate it. The buttons in the remote control provides functionality; the functionality of buttons in the remote control is implemented in the television set. The remote control just provides an interface of the functionality of the television. This is called *abstraction*. Another example, the table of contents of a book is an abstraction and the order of topics listed in the table of contents written inside the book is implementation. It may happen that two books have the same table of contents but the way it is represented may be different.

13.2 ABSTRACT CLASS AND ABSTRACT METHOD

Abstract class is a template for creating a concrete child class. It allows us to create concrete methods that are implemented in the child class. An abstract method is a method that has a declaration but no implementation. A class that has at least one abstract method is known as an abstract class. All the child classes must implement the abstract methods defined in the parent class.

DOI: 10.1201/9781003219125-15

Example 13.1

```
class A:
  def method_a(self):
    pass

class B(A):
  def method_a(self):
    print("method_a implemented")
obj_a = A()
obj_a.method_a()

obj_b = B()
obj_b.method_a()
```

In this example *class A* is a base and *class B* is the derived class. There is a method *method_a* in both classes. But in class A, *method_a* is only the declared, as there is no body for this method (implementation of the method is not defined). And in class B this method is declared and defined as well. We can instantiate both the class *Obj_a* and *Obj_b* there is no error because both classes are concrete classes. It is just single-level inheritance.

In Python there is no direct method for implementing the concept of abstract class. In order to implement an abstract class, the following steps must be followed:

1. Import two functions *ABC* and *abstractmethod* from the built-in module *AbstractBaseClass (abc)*.
2. The parent class must inherit from the base class *ABC*.
3. Use the decorator *@abstractmethod* before all those methods which are abstract methods in the base class.
4. The body of the abstract methods must not have implementation.
5. Implement the abstract methods in the child class.

The first three steps are a must to make a class abstract class. The fourth step is the implementation of the abstract method. We will get an error if any of these steps are missed. In inheritance, we can create the instances of the parent class as well as the child class but in the abstract class we cannot create instances of the parent class.

Example 13.2

```
from abc import ABC, abstractmethod
class A(ABC):
  @abstractmethod
  def method_a(self):
    pass
```

```
class B(A):
  def method_a(self):
    print("method_a implemented")

obj_b = B()
obj_b.method_a()

obj_a= A()
obj_a.method_a()
```

Observe the 1st line of the code, we have imported *ABC* and *abstractmethod* from the module *abc*. In the 2nd line we have created an abstract class *A* and this class inherits from *ABC*. In the 3rd line, we have used a decorator *@abstractmethod* before the method declaration as a result *method_a* is treated as an abstract method.

In the 6th line we have class B inherit class A. The next line is *def method_a(self):* (abstract method in class A), which is implemented in class B. In the 9th line, we have created an object of *class B* and in the next line we called *method_a* of *class B*. Observe the output, it prints *method_a implemented*. But when we tried to create an object of *class A*, it raised an error *Can't instantiate abstract class A with abstract methods method_a*. The error is because of the fact that abstract class cannot be instantiated.

Output

```
method_a implemented
Traceback (most recent call last):
  File "C:\Users\RakeshN\AppData\Local\Programs\Python\Py
  thon37-32\chap-13.py", line 31, in <module>
  obj_a= A()
TypeError: Can't instantiate abstract class A with abstract
methods method_a
```

Example 13.3

```
from abc import ABC,abstractmethod
class Automobile(ABC):
  @abstractmethod
  def move(self):
    pass
class Bike(Automobile):
  def move(self):
    print("Land Vehicle")

class Yatch(Automobile):
  def move(self):
    print("Water Vehicle")
```

```
b1 = Bike()
b1.move()

y1 = Yatch()
y1.move()
```

In this example, in the *Automobile* class, as *move()* is an abstract method, the *Automobile* class is also an abstract class. The abstract method *move()* is implemented in class (*Bike* and *Yatch*) which inherits the class *Automobile*.

Output

```
Land Vehicle
Water Vehicle
```

An interface is a set of publicly accessible methods on an object which can be used by other parts of the program to interact with that object. In Python the concept of interface is irrelevant. As Python supports multiple-inheritance it does not need an interface.

13.3   RELATIONSHIP

Objects cannot stand alone. In the real world, objects of one class have some connection with objects of another class. This connection is nothing but relational. For example, a user has a password, a university has many departments, a department has one or more teachers, a teacher teaches many subjects, students opt for different subjects, etc. In all these examples there is a relationship between objects of different classes. Again, these relationships can be:

- *One-to-One*: the relationship between user and password. Every user has only one password.
- *One-to-Many*: the relationship between manager and employee. There are many employees working under the supervision of one manager.
- *Many-to-One*: The relationship between chapters and a book. There are many chapters in a book.
- *Many-to-Many*: the relationship between student and subjects. There are many students who opt for many subjects.

13.3.1   ASSOCIATION RELATIONSHIP

The association relationship is also known as *has a* relationship. Some examples of association are: an employee *has a* name, an employee *has an* employee number. When we create an object of a class, by default all the attributes defined inside it pose as a *has a* relationship with it. An association relationship can be there between two different objects where these objects are not dependent on one another and they can have their own lifetime. They come together to establish a relationship.

An example of association is between a person and a laptop. A person has its own attributes and a laptop also has its own. The relationship between them is *person has a laptop*. Here person owns a laptop.

FIGURE 13.1   Association relationship

Another example of association is between an employee and a card. An employee has a card to enter the office premise. If a card is issued to an employee, we can find the card details if we know the employee's details and vice-versa. The association relationship is denoted by a line connecting both the classes as shown in figure 13.1.

In order to implement association in Python, the following steps must be followed:

1. Create two concrete classes (say class A and class B).
2. Create the objects in both classes.
3. Pass the object of one class as a parameter to the method of another class.

Example 13.4

```
class Employee:
  def __init__(self,id,name,age):
    self.id= id
    self.name = name
    self.age = age

  def login(self, Obj):
    return Obj.no
  def __str__(self):
    return "Emp Id" + str(self.id) + " has Name "+self.name
    + " is " + str(self.age)

class Card:
  def __init__(self, no, typ, exp):
    self.no = no
    self.type = typ
    self.exp = exp

  def swipe(self, Obj):
    return Obj.name
    def __str__(self):
    return "card no "+ self.no+" is of type " + self.type +
    " Expires on "+ self.exp

ObjEmp = Employee(501,'Sushobhan',26)
print(ObjEmp)
ObjCard = Card("1-00-1-9","RIFD","12-08-2021")
print(ObjCard)
print( ObjCard.swipe(ObjEmp),' uses card no ', ObjEmp.login(
ObjCard))
```

Observe that we are having two classes *Employee* and *Card* with objects *ObjEmp* and *ObjCard* respectively. The Employee class has a *login()* method that takes the card object while the *swipe()* method in Card class takes the object of Employee class.

Output

```
Emp Id501 has Name Sushobhan is 26
card no 1-00-1-9 is of type RIFD Expires on 12-08-2021
Sushobhan uses card no 1-00-1-9
```

There are association relationships which can be visualized as *whole-part* relation-ships. The combination of *parts* is a *whole*. For example, in a *tree and leaf relation, leaf is a part and tree is a whole.* Similarly, in a *computer system and keyboard* relation, the keyboard is a *part* and the computer system is a *whole*. Depending on the situation the *part* may/may not exist when the *whole* does not exist. As shown in figure 13.2, we can categorize the concept of association into two types based on the existence of *part* when the *whole* does not exist.

1. Aggregation relation
2. Composition relation

The composition is a stronger form of aggregation relation and aggregation is a stron-ger form of association.

13.3.2 AGGREGATION (USES A) RELATIONSHIP

Aggregation is a special form of association relationship. An example of aggregation is the relation between a player and a team. Generally, players are used to form a team. As long as they play together, they are a team. In other words, the team owns the players. When the team dissolves, the player can be a part of another team.

Another example is the relationship between a computer system and the compo-nents of a computer system. A computer system uses all the components together for proper functioning. A mouse is a component of a computer system that can be removed from one computer and can be used with another computer. It means a computer system uses a mouse. As long as a mouse is connected to the system, the system owns the mouse. This type of relationship is known as *uses a* relationship.

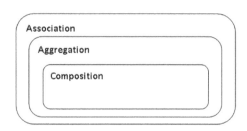

FIGURE 13.2   Association and its types

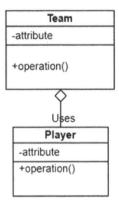

FIGURE 13.3 Aggregation relationship

In this *whole-part* relationship the *part* can exist without the *whole*. In other words, if the container object is deleted, its content objects may still exist. This relationship is denoted by a hollow diamond head towards the *whole* as shown in figure 13.3.

As we have seen that aggregation is a type of association, the object of one class uses the object of another class. The object which uses another object is the *container* and the object which is used is the *content*.

Aggregation is a stylish way of implementing inheritance. In inheritance, by default all the public attributes and methods are inherited. It becomes complicated as the level of inheritance increases. Whereas in aggregation, we take only those methods which we needed from the class defined in the parent class. In other words, with aggregation we can explicitly choose what to choose from the other class. It also prevents the complicated *Diamond Problem* in inheritance.

In order to implement aggregation in Python, the following steps must be followed:

1. Create two concrete classes, one class as a container (say class A) and the other as content (say class B).
2. Create the objects of content (class B) first and then create the object of the container (class A).
3. Pass the object of content (object of class B) as an argument to a method of container (object of class A).
4. Now assign the object of content (object of class B) passed to a member variable of container (object of class A).
5. We can now call any public method defined inside the content (class B).

Let us try to understand the above steps with the help of a simple example.

Example 13.5

```
class B:
  def some_method_in_B(self):
```

```
    print("Inside class B, some_method()")

class A:
  def __init__(self, obj_B):
    print("Initializer of class A")
    self.Obj_B_inside_A = obj_B

  def access_method_of_B(self):
    self.Obj_B_inside_A.some_method_in_B()

print("create object of B")
print("_"* 20)
obj_B = B()
obj_B.some_method_in_B()

print("create object of A")
print("_"* 20)
obj_A = A(obj_B)
obj_A.access_method_of_B()
```

We can see that there are two classes defined Class B and Class A. Class B is a simple concrete class. Class A is also a concrete class but the object of class B (obj_B) is passed as an argument to the *__init__()* method while defining class A. The object of class B (obj_B) is received as a parameter in the *__init__()* method. The line *self.Obj_B_inside_A = obj_B* enables this object used as any simple attribute of class A. Now we can use any public method defined inside class B by *self.object_name.method_name()*, in this case *self.Obj_B_inside_A. some_method_in_B()*.

Output

```
create object of B
_____
Inside class B, some_method()
create object of A
_____
Initializer of class A
Inside class B, some_method()
```

The following code establishes a relationship between Department and Faculty. The relation is *Department uses the services of Faculty*. The Faculty still survive even if the Department does not exist.

Example 13.6

```
class Faculty:
  def __init__(self,Fid,Fname):
    self.Fid = Fid
    self.Fname = Fname
```

```
    def teaches(self,subject):
      self.subject = subject
      print("Faculty ",self.Fname," Teaches ",self.subject)

class Department:
  def __init__(self,Dname,ObjFaculty):
    self.Dname = Dname
    self.faculty = ObjFaculty

ObjTeacher = Faculty(507,'Rajesh')
ObjDept = Department('Computer Science',ObjTeacher)
ObjDept.faculty.teaches('Data Structures')
```

Here the object of the *Department* class is the container and the object of the *Faculty* class is content. After creating the object of *Faculty* that is ObjTeacher = Faculty(507,'Rajesh'), it is passed as an argument to the constructor of the *Department*. This (*ObjTeacher*) is treated as any other parameter and is assigned to a variable. Now all the public methods of class *Faculty* is accessible. In the statement *ObjDept.faculty.teaches('Data Structures')*, the *teaches()* method defined inside *Faculty* is accessed from the object of *Department*.

Output

```
Faculty Rajesh Teaches Data Structures
```

We can observe that the aggregation relation is a unidirectional association. It means the relationship is from container to content, not vice-versa. Both objects are independent of each other, both objects can survive individually.

13.3.3 COMPOSITION (HAS A) RELATIONSHIP

Composition is a special form (strong form) of aggregation. Delegating the responsibilities of one class to another class is called *Composition*. An example of composition is the relation between *a book* and *chapters in the book*. Neither a book is a chapter, nor a chapter a book. So, we cannot use inheritance. Similarly, neither a book uses a chapter, nor the chapter uses the book. So, we cannot use aggregation. The relation between book and chapter is *a book has chapters*. Chapters can never exist without a book. When the book will be destroyed the chapters will also be destroyed.

Another example is the relation between the parts of a human body and the human body. The parts of a human body cannot exist without the human body. So, we can say a human body has human body parts.

This type of relationship is known as *has a* or *part of* relationship. In this *whole-part* relationship, the *part* cannot exist without the *whole*. In other words, if the container object is deleted, its entire content is deleted as well.

As we have seen that composition is a type of association, the object of one class uses the object of another class. The object which uses another object is the *container*

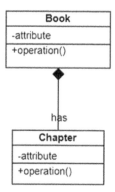

FIGURE 13.4   Composition relationship

and the object which is used is the *content*. This relationship is denoted by a solid diamond head towards the *whole* as shown in figure 13.4.

Composition is a stylish way of implementing inheritance. In inheritance, by default all the attributes and methods are inherited. It becomes complicated as the level of inheritance increases or in hybrid inheritance, whereas in composition we take only those methods which we needed from the class defined earlier. In other words, with the composition we can explicitly choose what to choose from the parent class.

In order to implement composition in Python, the following steps must be followed:

1. Create two concrete classes, one class as a container (say class A) and the other as content (say class B).
2. Create the objects of content (class B) inside the object of the container (class A).
3. Pass the arguments (if any) of the content (class B), as well as the arguments of container (class A) to the, *init__(self)* function of class A.
4. We can now call any public method defined inside the content (class B).

Let us try to understand the above steps with the help of a simple example.

Example 13.7

```
class B:
  def some_method_in_B(self):
    print("Inside class B, Some Method()")

class A:
  def __init__(self):
    print("Initializer for class A")
    self.Obj_B_inside_A = B()
```

```
    def access_method_of_B(self):
      self.Obj_B_inside_A.some_method_in_B()

print("\n create object of A")
print("-" * 20)
obj_A = A()
obj_A.access_method_of_B()
```

We can see that there are two classes defined Class B and Class A. Class A and B are simple concrete classes. Observe *self.Obj_B_inside_A = B()* inside the __ *init__()* of class A. It means an object of class B is created with the name *Obj_B_ inside_A* when the object of class A is created. Now we can use any public method defined inside class B by *self.object_name.method_name()*, in this case *self.Obj_B_inside_A.some_method_in_B()*. When the object of class A is deleted, the object of class B is also deleted.

Output

```
create object of A
--------------------
Initializer for class A
Inside class B, Some Method()
```

The following code establishes a relation between student and marks. The relation is *student has marks*. Marks have a meaning only when it is associated with any student. There is no meaning of the mark object without a student object. So the object mark is created inside the student object.

Example 13.8

```
class marks:
  def __init__(self,m1,m2,m3):
    self.mark1 = m1
    self.mark2 = m2
    self.mark3 = m3

  def total_mark(self):
    return self.mark1 + self.mark2 + self.mark3

class student:
  def __init__(self,roll,name,gender,m1,m2,m3):
    self.roll = roll
    self.name = name
    self.gender = gender
    self.mark = marks(m1, m2, m3)

  def avg(self):
    return self.mark.total_mark()/3
```

```
stud_obj = student('17UKA1042','Rajesh','Male',56,75,78)
print("The Average mark is ",stud_obj.avg())
```

Here the object of *student* class is the container and the object of *marks* class is content. The *__init__()* of *marks* takes three parameters m1, m2 and m3. As we are creating the object of *marks* inside *student,* we need to pass all the arguments that are required for *student* and *marks* to the *__init__()* of class *student.* The statement *self.mark = marks(m1, m2, m3)* creates an object of class marks, with object name *mark.*

Now all the public methods of class *marks* are accessible. In the statement *return self.mark.total_mark()/3, self.mark* is the object name and *total_mark()* is the method defined in the class marks.

Output

```
The Average mark is 69.66666666666667
```

13.4   DIFFERENCE BETWEEN INHERITANCE AND ASSOCIATION

Inheritance	Association
A class is inherited from another class	Association uses an object of a class
There is a concept of parent and child class	There is no such concept of parent and child class
There is no such concept of container and content	There is a concept of container and content
All the public attributes and methods of parent class are available to child class	Selective methods and attributes are available
It refers to *is a* relationship	It refers to *has a* or *uses a* relationship

13.5   DIFFERENCE BETWEEN AGGREGATION AND COMPOSITION

Composition	Aggregation
Composition is a strong form of aggregation	Aggregation is a weak form of composition
Content object cannot exist without container object	Content object can exist without container object
It is a unidirectional relation	It is a unidirectional relation
It refers to *has a* relationship between objects	It refers to *uses a* relationship between objects

WORKED OUT EXAMPLES

Example 13.9 Write a Python program that implements aggregation relation Employee *has Salary.*

```
class Salary:
  def __init__(self,basic,da,hra,bonus):
```

```
    self.basic = basic
    self.da = da
    self.hra = hra
    self.bonus = bonus

  def Pay(self):
    return self.basic + (self.basic *self.da / 100) + self.h
    ra + self.bonus

  def Annual_Salary(self):
    return self.Pay()*12

class Employee:
  def __init__(self,id,name,sal):
    self.id = id
    self.name = name
    self.obj_sal = sal

  def Emp_Salary_Monthly(self):
    return self.obj_sal.Pay()

  def Emp_Salary_Annualy(self):
    return self.obj_sal.Annual_Salary()

sal = Salary(22000,10,1200,0)
emp = Employee('501','Nakshatra',sal)
print(emp.name," Monthly Salary ",emp.Emp_Salary_Monthly())
print(emp.name," Annual Salary ",emp.Emp_Salary_Annualy())
```

Output

```
Nakshatra Monthly Salary 25400.0
Nakshatra Annual Salary 304800.0
```

Example 13.10 Write a python program that implements the composition relation between Employee and Salary

```
class Salary:
  def __init__(self,basic,da,hra,bonus):
    self.basic = basic
    self.da = da
    self.hra = hra
    self.bonus = bonus

  def Pay(self):
    return self.basic + (self.basic * self.da / 100) +
    self.hra + self.bonus
```

```
    def Annual_Salary(self):
      return self.Pay()*12

class Employee:
  def __init__(self,id,name, basic,da,hra,bonus):
    self.id = id
    self.name = name
    self.obj_sal = Salary(basic,da,hra,bonus)

  def Emp_Salary_Monthly(self):
    return self.obj_sal.Pay()

  def Emp_Salary_Annualy(self):
    return self.obj_sal.Annual_Salary()

emp = Employee('501','Nakshatra',22000,10,1200,0)
print(emp.name," Monthly Salary ",emp.Emp_Salary_Monthly())
print(emp.name," Annual Salary ",emp.Emp_Salary_Annualy())
```

Output

```
Nakshatra Monthly Salary 25400.0
Nakshatra Annual Salary 304800.0
```

MULTIPLE CHOICE QUESTIONS

1. What relationship correctly fits for university and professor?
   a. Aggregation
   b. Composition
   c. Inheritance
   d. All of the above

2. What relationship is best suited for house and door?
   a. Association
   b. Composition
   c. Inheritance
   d. All of the above

3. What relationship is best suited for employee and person?
   a. Association
   b. Composition
   c. Inheritance
   d. All of the above

4. In inheritance
   a. Only a few public methods/attributes are accessed by child class
   b. Only private methods/attributes are accessible by child class
   c. All public methods/attributes are accessible by child class
   d. All of the above

5. In aggregation
   a. Only a few public methods/attributes are accessed by child class
   b. Only private methods/attributes are accessible by child class
   c. Only public methods/attributes are accessible by child class
   d. All of the above

6. The concept of interface is irrelevant as Python supports
   a. Hybrid Inheritance
   b. Multiple Inheritance
   c. Single-level inheritance
   d. Multi-level inheritance

7. Which relationship is properly described by *has a*
   a. Aggregation
   b. Composition
   c. Inheritance
   d. All of the above

8. Which relationship is properly described by *part of*
   a. Aggregation
   b. Composition
   c. Inheritance
   d. All of the above

9. Which relationship is properly described by *is af*
   a. Aggregation
   b. Composition
   c. Inheritance
   d. All of the above

10. Which relationship is properly described by *is a*
    a. Aggregation
    b. Composition
    c. Inheritance
    d. All of the above

11. The container is a
    a. Class
    b. Object
    c. Relation
    d. All of the above

12. The content is instantiated inside the container in
    a. Aggregation
    b. Composition
    c. Inheritance
    d. All of the above

13. The content is instantiated outside the container in
    a. Aggregation
    b. Composition
    c. Inheritance
    d. All of the above

14. The concept of content and container is not in
    a. Aggregation
    b. Composition
    c. Inheritance
    d. All of the above

15. Which of the following is/are NOT true?
    a. Aggregation is a weak form of composition
    b. Aggregation is a special form of association
    c. Inheritance refers to *is a* relation
    d. All of the above are true

16. The following code refers to
    ```
    class B:
        def some_method_in_B(self):
            print("Inside class B, some_method()")
    class A:
        def __init__(self):
            print("Initializer of Class A")
            self.Obj_B_inside_A = B()
    ```

    a. Aggregation
    b. Composition
    c. Inheritance
    d. None of the above

17. Which module is to be imported for abstract class
    a. abc
    b. ABC
    c. @abstractclass
    d. None of the above

18. A class is an abstract class when it has?
    a. The body of the method may contain a pass
    b. At least one method is abstract
    c. For each abstract method the decorator @abstractclass is used
    d. The abstract class must inherit from ABC

19. Implementation of abstract class is done at
    a. The place where it is defined
    b. The place where it is invoked
    c. In the child class
    d. All of the above

20. While instantiating abstract class, we get
    a. Attribute Error
    b. TypeError
    c. SyntaxError
    d. ValueError

DESCRIPTIVE QUESTIONS

1. What is an association relationship? Explain with an example.
2. What are the different forms of association relationships? Explain with an example.
3. What is a composition relation? Explain with an example.
4. What is Aggregation relation? Explain with an example.
5. Differentiate between different types of relations.

PROGRAMMING QUESTIONS

1. Write a Python program in which there are two classes Driver and Car. Show the *association* relationship between them that shows which driver is driving which car.
2. Write a Python program in which there are two classes Person and Address. Show *aggregation* relation between them with *person has an address*.
3. Write a Python program in which there are two classes Book and Chapters. Show *composition* relation between them with *Book has Chapters*.
4. Write a Python program that creates an abstract class Person with an abstract method *work_of()*. The concrete class Student and Teacher implement the *work_of()* method.

ANSWER TO MULTIPLE CHOICE QUESTIONS

1.	A	2.	B	3.	C	4.	C	5.	A
6.	D	7.	B	8.	A	9.	C	10.	C
11.	B	12.	C	13.	A	14.	C	15.	D
16.	B	17.	A	18.	A	19.	C	20.	B

14 Exception Handling

LEARNING OBJECTIVES

After studying this chapter, the reader will be able to:

- Difference between exception and error
- Know how to use an exception
- Know about exception hierarchy
- Define and use an exception in a function
- Raise an exception
- Define and use a user-defined exception
- Know assertion

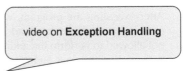
video on **Exception Handling**

14.1 ERRORS AND EXCEPTIONS

When writing a program, we come across different types of errors and at different stages. The errors can be:

- Syntax errors.
- Semantic errors.
- Logical errors.

Syntax refers to the rules and grammar of a language. We get these errors if we don't follow the construct or make a spelling mistake, improper indentation or improper arguments. For example, while writing *if-else* statement, if we forget to use a colon (:) after condition or in place of keyword *while*, if we write *wihle*. These errors are reported during compilation. Syntax errors, also known as parsing errors, are the most common kind of errors. And they occur when we do not follow the grammar of the language. Let us try this example in Python prompt.

Example 14.1

```
>>> print ("I am Learning Python"
```

Output

```
SyntaxError: Missing parentheses in call to 'print'. Did you
mean print("I am Learning Python)?
```

Observe the error message. It is a syntax error as clearly seen that the right parentheses are missing. Python parser shows the line number, in which the error appeared. Apart from that it also displays information about the type and position of the error.

Semantic refers to the meaning of a statement. We get these errors if the rules of the grammar are correct but it does not provide any meaning. For example, the sentence "Kite is flying high in the sky" is syntactically correct, because it follows the English language grammar rules. It is also semantically correct because it has a meaning. However, the sentence "Elephant is flying high in the sky", is syntactically correct, but it is semantically incorrect because it has no meaning. These errors are reported during execution or run-time that disrupts the flow of program instruction.

Example 14.2

```
x = 5
y = 'cat'
print("The sum is ", x+y)
print("The program ends here")
```

After executing the program, we can see there is an error.. The reason for the error is that when we try to add x and y, nothing wrong with the syntax, but it is not allowed to perform the addition operation with an operand with two different datatypes (in this case an Integer and a String). So, we get a *TypeError*.

Output

```
Traceback (most recent call last):
  File"C:/Users/RakeshN/AppData/Local/Programs/Python/Pytho
  n37-32/chap-14 Exception Handling.py", line 6, in <module>
  print("The sum is ", x+y)
TypeError: unsupported operand type(s) for +: 'int' and 'str'
```

The third type of error is the logical error, it refers to the logic of the program. These errors are difficult to identify as it will not raise any errors. Even though the inputs are correct, the outputs are wrong. Identifying a statement where logical error occurred is difficult. Generally, we unfold these errors while testing the program.

14.2   THE TRY:, EXCEPT:, ELSE: AND FINALLY: BLOCK

Out of the errors we discussed in the previous section, the runtime errors occur are the errors that occur based on the input given by the user. In the above example (example-14.2) there is no error if the value of y is any numeric value. The errors done by the user while providing data for an operation can be handled. We can catch those errors and handle it ourselves is known as *Exception Handling*. This will not disrupt the flow of the program. The general syntax is:

try:
        statement(s)
except (exception_1,var) :
        statement(s)
except (exception_2,var) :
        statement(s)

else:
    statement(s)
finally: ,
    statement(s)

In the above syntax notice that:

- The statement(s) between *try* and *except* keywords are executed and checked for any errors. The statements that may raise an error are kept inside the *try*-block.
- *except* is the keyword followed by the optional *exception_1, var* within a pair of parentheses followed by a colon.
- At least one *except* clause is desirable to handle the exception in the structure.
- The optional *else* clause comes after all except clauses. It is useful for code that must be executed if the **try** clause does not raise an exception.
- If no exception occurs, then the *except* clause is skipped and the statements in the *else* are executed.
- If an exception occurs, then the rest of the clauses are skipped and the corresponding *except* clause is executed.
- The *except* clause with no exceptions handles any kind of exception.
- If an optional *finally* clause exists, the statements in the *finally* clause will always be executed irrespective of any exception. Usually, the purpose of the *finally* clause is to clean up your code if it is needed.
- Here *exception_1, exception_2, else* and *finally* are the keywords that are optional. The variable *var* is also optional.
- *exception_1* and *exception_2* are descriptions of the exception like FileNotFoundError, NameError, etc.

Let's see some examples with different kinds of exceptions.

14.2.1   THE TRY: AND EXCEPT: BLOCK

The first example is a *TypeError* exception. This exception happens when we try to operate on operands of different datatypes which is not allowed.

Example 14.3

```
x = 5
y = 'cat'
try:
  print("The sum is ",x + y)
except:
  print("Not Allowed to add operand of different data types")
print("The program ends here")
```

When a number is added to a String, it raises an exception *TypeError*, we can handle it by writing the statement that may cause an error inside the *try*-block and an appropriate print statement inside the *except*-block. We can also observe that the last print statement also got printed even though there is an error as this print statement is outside any block. In other words, the flow of the program is not disrupted.

Output

```
Not Allowed to add operand of different data types
The program ends here
```

Example 14.4 Zero division error

```
x = 5
y = 0
try:
  print(x/y)
except:
  print("Operation Not allowed...")
```

ZeroDivisionError occurs when a number is divided by zero. Here the critical statement that may raise an error is *print(x / y)*. So, this statement is kept inside the *try*-block. Whenever there is an error raised, the *except*-block is executed.

Output

```
Operation Not allowed...
```

14.2.2 THE TRY:, EXCEPT: AND ELSE: BLOCK

The *try-except* statement has an optional *else*-clause, which must come after all *except*-clauses. It is useful for code that must be executed if the *try*-clause does not raise any exception.

Example 14.5

```
try:
  file_pointer = open("MyTest_File.txt","r")
except:
  print("Could not Open File...")
else:
  x = file_pointer.read()
  print(x)
  file_pointer.close()
```

Observe that in the *try*-clause we are trying to open a file *"MyTest_File.txt"* that does not exist. As a result, it raised an exception. That exception is handled by *except*-block.

Output

```
Could not Open File...
```

Notice that the *else*-clause will be executed only when there is no exception occurs. It means if there is no error while opening the file "*MyTest_File.txt*", it will execute the *else*-block. When there is no error while opening the file, we can read the file and print the data.

Example 14.6 IOError exception

```
try:
  file_pointer = open("MyTestFile.txt","r")
except:
  print("Could not Open File...")
else:
  x = file_pointer.read()
  print(x)
  file_pointer.close()
```

Output

```
I am Learning Python
I am Enjoying it
```

We have already mentioned that the *else*-clause is optional. All the statements after *except*-block can be written with proper indentation.

Example 14.7 IOError exception

```
try:
  file_pointer = open("MyTest_File.txt","r")
except:
  print("Could not Open File...")
x = file_pointer.read()
print(x)
file_pointer.close()
```

But if an exception is raised, even then it will try to execute the last three lines. As the scope of variable file pointer is limited only to *try* and *except* statement. We have used it outside the *try-except* clause so we get *NameError*. So, it is always advisable to use the else clause.

Output

```
Could not Open File...
Traceback (most recent call last):
File "C:/Users/RakeshN/AppData/Local/Programs/Python/Python
37-32/chap-14 Exception Handling.py", line 61, in <module>
```

```
  x = file_pointer.read()
NameError: name 'file_pointer' is not defined
```

14.2.3 THE TRY:, EXCEPT :, ELSE: AND FINALLY: BLOCK

We can add an optional *finally* clause at the end of the *try-except* clause. The *finally* clause will always be executed no matter if we have an exception or not. Usually, the purpose of the finally clause is to clean up our code if it is needed.

Example 14.8 Finally clause

```
try:
  file_pointer = open("MyTest_File.txt","r")
except:
  print("Could not Open File...")
else:
  x = file_pointer.read()
  print(x)
  file_pointer.close()
finally:
  print("If any other Error... or Done...")
```

Observe that there is an exception raised and the *finally* clause also got executed.

Output

```
Could not Open File...
If any other Error... or Done...
```

Example 14.9 Finally clause

```
try:
  file_pointer = open("MyTestFile.txt","r")
except:
  print("Could not Open File...")
else:
  x = file_pointer.read()
  print(x)
  file_pointer.close()
finally:
  print("If any other Error... or Done...")
```

Observe that there is no exception raised yet still the finally clause was executed.

Output

```
I am Learning Python
I am Enjoying it
If any other Error... or Done...
```

It means the *finally* clause will be executed when all *exception* clauses and *else* clauses are executed.

14.2.4   MENTIONING THE ERROR DESCRIPTION

Instead of only writing *except*, we can also mention the error description within a pair of parentheses after the keyword except followed by a colon.

Example 14.10 Exception handling

```
try:
  file_pointer = open("MyTest_File.txt","r")
except:
  print("File Not Found...")
else:
  x = file_pointer.read()
  print(x)
  file_pointer.close()
```

Observe that we have mentioned the exact error name (which is pre-defined) to handle the error.

Output

```
File Not Found...
```

We can also write more than one pre-defined error name inside the parenthesis separated by a comma. We can handle all these errors with the same message.

Example 14.11 Exception handling with predefined error name

```
try:
  file_pointer = open("MyTest_File.txt","r")
except(FileNotFoundError, NameError):
  print("There is an Error...File Not Found...")
else:
  x = file_pointer.read()
  print(x)
  file_pointer.close()
```

Observe that we have written two pre-defined error names within the bracket. When either of the errors *FileNotFoundError* or *NameError* occurs, we get the same error message *There is an Error.*

Output

```
There is an Error...File Not Found...
```

We have already described that writing the name of the pre-defined error name in except clause is optional. Even though we don't mention the error name, still it will work as we have shown in earlier example.

It is also possible to do aliasing. Aliasing is a technique by which instead of writing the full error name we can assign a short name and refer to it.

Example 14.12 Exception aliasing

```
x = 5
y = 'cat'
try:
  z = a + y
except (NameError,TypeError) as e:
  print(e)
```

Here for the exception name *NameError* and *TypeError* we have an alias *e*. When we printed *e*, only the exception is printed. It will not print anything regarding the filename, line numbers or the statement where the error occurred.

Output

```
name 'a' is not defined
```

14.3   THE EXCEPTION HIERARCHY

It is worth mentioning that there are 63 exceptions, and all of them form a tree-shaped hierarchy in Python. Figure 14.1 shows a few of them. For a complete list we can refer to Python documentation in www.python.org/doc/.

Some of the built-in exceptions are more general while others are completely concrete. The exceptions located at the branches' end (leaf) are concrete; whereas the exceptions closer to the root are more general.

Let's begin examining the tree from the *ZeroDivisionError* at the leaf of the hierarchy.

- *ZeroDivisionError* is a special case of more a general exception class named *ArithmeticError*.

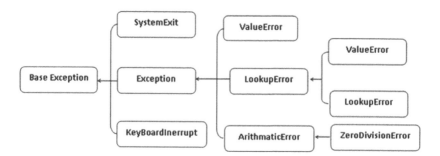

FIGURE 14.1   Exception hierarchy

- *ArithmeticError* is a special case of a more general exception class named *Exception*.
- *Exception* is a special case of a more general class named *BaseException*.

It is possible to replace *ZeroDivisionError* with *ArithmeticError*. We already know that *ArithmeticError* is a general class including (among others) the *ZeroDivisionError* exception.

Example 14.13 Exception handling with specific error name

```
x = 4
y = 0
try:
  z = x/y
except (ZeroDivisionError):
  print("Zero is in the Denominator...")
```

Observe that in this code we have used the error name *ZeroDivisionError* and the message printed as defined.

Output

```
Zero is in the Denominator...
```

The code's output remains unchanged even if we change the exception name to *ArithmeticError*. It implies that replacing the exception's name with either *Exception* or *BaseException* won't change the program's behavior.

Example 14.14 Exception handling with general error name

```
x = 4
y = 0
try:
  z = x/y
except (ArithmeticError):
  print("Zero is in the Denominator...")
```

The order of the branches matters. Don't put more general exceptions before more concrete ones. This will make the latter one unreachable and useless. Moreover, it will make our code messy and inconsistent. Python won't generate any error messages regarding this issue.

We already know that *ZeroDivisionError* is more specific and *ArithmeticError* is a more general error. When we use the more specific error name before a more general error name in the same branch, it will handle the error first.

Example 14.15 Exception handling

```
x = 4
y = 0
try:
  z = x/y
except (ZeroDivisionError):
  print("ZeroDivisionError...Zero is in the Denominator...")

except (ArithmeticError):
  print("ArithmeticError...Zero is in the Denominator...")
```

Observe that here *ZeroDivisionError* is used before *ArithmeticError*. When division by zero occurs, it is handled by *ZeroDivisionError*.

Output

```
ZeroDivisionError...Zero is in the Denominator...
```

In the next example, a more general error name is used before the more specific error name in the same branch. The error that occurs will be handled by the general error name.

Example 14.16 Exception handling

```
x = 4
y = 0
try:
  z = x/y

except (ArithmeticError):
  print("ArithmeticError...Zero is in the Denominator...")

except (ZeroDivisionError):
  print("ZeroDivisionError...Zero is in the Denominator...")
```

Observe that here *ArithmeticError* is used before *ZeroDivisionError*. When division by Zero occurred; it is handled by *ArithmeticError*.

Output

```
ArithmeticError...Zero is in the Denominator...
```

We can also use *try-except* block inside a looping structure.

Example 14.17

```
def EnterInteger():
  while True:
    try:
      i=int(input("\nEnter an Integer :"))
```

```
    finally:
    print("Finally, is executed...")
    try:
      z = 3 / i
    except:
      print("It is NOT an Integer, please enter an Integer")
      continue
    else:
      return(i)
    finally:
    print("Finally, is Executed...\n")
print("The Integer entered is ",EnterInteger())
```

Here *EnterInteger()* is a simple function that accepts an Integer. The *input()* statement is kept inside the first *try*-clause of *while*-loop. This will make the user enter an Integer until he supplies it. Each time the user supplies an input (valid or invalid), every time the *finally* clause is executed.

Output

```
Enter an Integer :we
Finally, is executed...
It is NOT an Integer, please enter an Integer
Finally, is Executed...
Enter an Integer :123
Finally, is Executed...
Finally, is Executed...
The Integer entered is 123
```

In section 14.2, we have discussed that it is desirable to have at least one *except* clause in a *try-except* block as we don't know how to handle if any error occurs in the *try*-clause. We can write a *finally*-clause immediately after a *try*-clause, it will not raise any syntax error but if an error occurs while executing the statements inside the *try-clause* it will show an appropriate error message after executing the statements inside the *finally*-clause.

Example 14.18

```
try:
  i = int(input("\nEnter an Integer : "))
finally:
  print("Finally is executed...")
```

Output 1

```
Enter an Integer : 5
Finally is executed...
```

Output 2

```
Enter an Integer : 5.3
Finally is executed...
Traceback (most recent call last):
File "C:/Users/RakeshN/AppData/Local/Programs/Python/Python3
7-32/chap-14 Exception Handling.py", line 193, in <module>
i = int(input("\nEnter an Integer : "))
ValueError: invalid literal for int() with base 10: '5.3'
```

It is possible to write another *try-except* clause inside *finally-block*. If a *try-except* clause doesn't have an *except*-clause, the error occurs in the *try*-block can be handled inside the *finally*-block by having another *try-except* block. We have modified example 14.17 to achieve the same output without *except*-clause.

Example 14.19

```
def EnterInteger():
  while True:
    try:
    i=int(input("\nEnter an Integer :"))

    finally:
      print("Finally, is executed...")
    try:
      z = 3 / i
    except:
      print("Please enter a Non-Zero Integer")
      continue
    else:
      return(i)

EnterInteger()
```

Observe that there is a *finally*-clause just after the *try-block*. Inside the *finally-block* another *try-except* block is present. This example shows there can be a *try-except* block inside the *finally*-block and can handle the error raised inside the *try*-block. The readers may try for a *try-except* clause inside the *except* clause.

Output

```
Enter an Integer :we
Finally, is exeuted...
Please enter a Non-Zero Integer

Enter an Integer :5.5
Finally, is executed...
Please enter a Non-Zero Integer

Enter an Integer :0
Finally, is executed...
```

```
Please enter a Non-Zero Integer

Enter an Integer :3
Finally, is executed...
```

14.4   EXCEPTION IN A FUNCTION

It is possible that an exception can be raised in a function. When an exception is raised inside a function, it can be handled either inside the function or outside the function. In order to handle the exception inside the function, the main statements (where we expect the error may raise), need to be kept inside the *try-except* block.

Example 14.20

```
def fun():
  x = 4
  y = 0
  try:
    z = x/y
  except (ZeroDivisionError):
    print("Zero Division Error...")

fun()
print("Program End...")
```

Observe that the *"ZeroDivisionError"* exception is raised inside the function *fun()*, and the function takes care of the error.

Output

```
Zero Division Error...
Program End...
```

It is also possible to let the exception propagate outside the function and then handle it with an appropriate message.

Example 14.21 Exception handling

```
def fun():
  x = 4
  y = 0
z = x/y
try:
  fun()
except (ZeroDivisionError):
  print("Zero Division Error...")

print("Program End...")
```

Observe that in this code the error handling is not done inside the function. Rather the error is handled after the function is called. The calling of the function is done inside a *try-except* block.

Output

```
Zero Division Error...
Program End...
```

14.5   THE RAISED EXCEPTION

There is one more way to handle exceptions that is by using *raised-exception*. The raised exception can traverse function and module boundaries, searching the invocation chain for a matching except clause to handle it.. If there is no such clause, the exception remains unhandled, and Python solves the problem in its standard way by terminating the code and emitting a diagnostic message.

The general syntax for raise is: ***raise [Exception Name [,argument]]exception***. In the above syntax notice that:

- Here *raise* is a keyword. This instruction raises a specific exception.
- Mentioning the *exception name* is optional.
- Not mentioning the Error Name refers to a general error.

The instruction enables us to:

- Simulate raising actual exceptions (e.g., to test our handling strategy)
- Partially handle an exception and make another part of the code responsible for completing the handling (separation of concerns).

There is one serious restriction:

- This kind of raise instruction may be used inside the except branch only.
- Using it in any other context causes an error.

Example 14.22 Exception handling with raise

```
def fun(n):
  x = 4
  return (x/n)
  raise error

try:
  print("The Result is :",fun(xxx))

except (ZeroDivisionError):
  print("The Denominator must be non-zero...")
except (TypeError):
```

```
  print("The Denominator must be number...")
except (NameError):
  print("The Denominator is NOT a valid number...")

print("Program End...")
```

Observe that the *raise error* appears inside the function. The actual exception handling statements are after the function call. Depending on the type of error specific error messages will be generated. In this example instead of a number in the denominator, if *xxx* (without any quotation mark) is passed as a String, a *NameError* is raised. Accordingly, a print statement was executed.

Output

> The Denominator is NOT a valid number...
> Program End...

Example 14.23 Exception handling with raise

```
def fun(n):
  x = 4
  return (x/n)
  raise Error

try:
  print("The Result is :",fun(0))

except (ZeroDivisionError):
  print("The Denominator must be non-Zero Number...")
except (TypeError):
  print("The Denominator must be number...")
except (NameError):
  print("The Denominator is NOT a valid number...")

print("Program End...")
```

In this example the same raise Error will invoke the *ZeroDivisionError* as the passed value denominator to the function is 0.

Output

```
The Denominator must be non-Zero Number...
Program End...
```

Example 14.24 Exception handling with raise

```
def fun(n):
  x = 4
  return (x/n)
  raise Error
```

```
try:
  print("The Result is :",fun('x'))

except (ZeroDivisionError):
  print("The Denominator must be non-Zero Number...")
except (TypeError):
  print("The Denominator must be number...")
except (NameError):
  print("The Denominator is NOT a valid number...")

print("Program End...")
```

When the input is 'x' (with quotation) the *raise error* will invoke *TypeError* as the value passed to the function is not a number but a String.

Output

```
The Denominator must be number...
Program End...
```

14.6   USER DEFINED EXCEPTIONS

The exception we have discussed in section 14.3 is neither closed nor complete. It means we can always extend it. Python allows users to define their own exceptions. In order to define a user-defined exception class, we need to derive from the built-in exception class. We need to define how to deal with the exception within the user-defined class. Let us try to understand by a simple program.

Example 14.25

```
class MyError(Exception):
  print("User Exception")

try:
  print("In the Try-Except Block")
  raise MyError
except MyError:
  print("Error Handled")
```

Here observe that *MyError* is a user-defined class that is inheriting from the class *Exception*. This exception can be raised using *raise* like any other exception.

Output

```
User Exception
In the Try-Except Block
Error Handled
```

In the next example, we are going to discuss user-defined exceptions in more detail. We are going to generate exceptions for the speed limit for a car manufactured by different companies. We can define a hierarchy of exceptions in user-defined exceptions also. The base exception in the hierarchy should always derive from any of the top exception classes of build-in Python exceptions such as *Exception*.

Example 14.26

```
class CarSpeedError(Exception):
  def __init__(self,car = 'unknown', message = ' Speed'):
  Exception.__init__(self, message)
  self.car = car

class TooMuchSpeedError(CarSpeedError):
  def __init__(self,car = 'unknown', speed ='>150',message =
  'Control Your Speed '):
  CarSpeedError.__init__(self,car,message)
  self.speed = speed

def carMake(car, speed):
  if car not in ['Jaguar', 'BMW', 'Honda']:
    raise CarSpeedError
  if speed >150 :
    raise TooMuchSpeedError
    print(Car,": car Speed under Control")

for (Car,Speed) in [('Jaguar',200),('BMW',120),('Honda',
100),('Matuti',80)]:
  try:
    carMake(Car,Speed)
  except TooMuchSpeedError as e:
    print(Car,":",e,":",e.speed)
  except CarSpeedError as e1:
    print(Car,":",e1,":",e1.car)
```

Observe that in this user-defined exception hierarchy, the base exception is *CarSpeedError* which is derived from *Exception*. It takes two parameters, *car* and *message*. The class *TooMuchSpeedError* is derived from *CarSpeedError*. So, a more specific *CarSpeedError* is *TooMuchSpeedError*. It takes three parameters. When the car is either 'Jaguar' or 'BMW' or 'Honda' and the speed is greater than 150, it raises more specific *TooMuchSpeedError* exception. If the car is not in the list, it raises a more general *CarSpeedError* exception.

Output

```
Jaguar : Control Your Speed : >150
BMW : car Speed under Control
Honda : car Speed under Control
Matuti : Speed : unknown
```

14.7   ASSERTION

Yet another way of handling error is *assertion*. We can think of this as a *raise if* statement. If the expression is evaluated as false, the assertion is raised.

The general syntax is: ***assert Expression[,Arguments]***.

The instruction notices that:

- The statement starts with the keyword *assert*.
- It is followed by an expression; which needs to be evaluated.
- If the expression evaluates to True, or a non-zero numerical value, or a non-empty String, or any other value different than None, it won't do anything else.
- Otherwise, it raises an exception named *AssertionError* (in this case, we say that the assertion has failed)

We may want to put it into our code where we want to be absolutely safe from evidently wrong data, and where we aren't absolutely sure that the data has been carefully examined before raising an *AssertionError* exception secures our code from producing invalid results, and clearly shows the nature of the failure. Assertions don't supersede exceptions or validate the data.

Example 14.27 Exception handling with assert

```
import math
x=float(input("Enter a positive number "))
assert x > 0.0, "You have Not entered a positive number"
print (math.sqrt(x))
```

In this example, we want the user to enter a positive number. It means whenever a user enters a negative number, an assertion should be raised. In 3rd line, "*assert x > 0.0*" indicates that whenever the value is less than or equal to 0.0, the assertion is raised.

Output

```
Enter a positive number -4
Traceback (most recent call last):
File "C:\Users\RakeshN\AppData\Local\Programs\Python\Python3
7-32\chap-14 Exception Handling.py", line 330, in <module>
assert x>0.0, "You have Not entered a positive number"
AssertionError: You have Not entered a positive number
```

WORKED OUT EXAMPLES

Example 14.28 Write a program that asks the user for an input and tries to handle the error that may occur when trying to type cast the input to an int using the *try ...except... else* **clause. The function should print the result if the operation is successful, if the operation is not successful it should print Error : error message.**

```
x = input ("Please enter something: ")
try:
  x = int (x)
except ValueError as e:
  print ('Error :',e)
else:
  print (x)
```

Output 1

```
Please enter something: h
Error : invalid literal for int () with base 10: 'h'
```

Output 2

```
Please enter something: 3
3
```

MULTIPLE CHOICE QUESTIONS

1. What is the output of the following code?
   a = 6
   b = 5
   z = (a+b/2a)

   a. Syntax Error
   b. Value Error
   c. Attribute Error
   d. 0.916

2. What is the output of the following code?
   a = 6
   b = 6 - 12
   z = a/(a+b)

   a. Syntax Error
   b. Value Error
   c. Attribute Error
   d. ZeroDivisionError

3. What is the output of the following code?

```
my_directory = {1:"One",
                2:"Two",
                3:"Three"}
print(my_directory["Two"])
```

   a. Syntax Error
   b. Value Error
   c. Key Error
   d. ZeroDivisionError

4. What is the output of the following code?

```
a = 505
b = 334234200
print(a**b)
```

   a. Syntax Error
   b. Overflow Error
   c. Key Error
   d. ZeroDivisionError

5. If there is a *finally*: block inside a *try*: block
   a. It will be executed when there is no else
   b. It will always be executed
   c. It will be executed sometimes
   d. It will be executed when no exception

6. What is the output of the following code?

```
try:
    raise Exception
except BaseException:
    print("Python",end = "")
else:
    print("Basic",end = "")
finally:
    print("Exception")
```

   a. Python
   b. PythonException
   c. Basic
   d. Basic Exception

7. What is the output of the following code?

```
try:
    raise Exception
except:
    print("Python")
except BaseException:
    print("Basic")
```

```
finally:
    print("Exception")
```

a. Python
b. Exception
c. Basic
d. Error

8. What is the output of the following code?
```
def f(x):
try:
    x = x/x
except:
    print("Python",end="")
else:
    print("Basic",end="")
finally:
    print("Exception")
f(1)
f(0)
```

a. PythonException and BasicException
b. BasicException and PythonException
c. PythonException
d. BasicException

9. What is the output of the following code?
```
try:
    raise Exception('a','b','c')
except Exception as e:
    print(e.args)
```

a. ('a','b','c')
b. ['a','b','c']
c. {'a','b','c'}
d. 'a','b','c'

10. The base class of built-in exceptions is
    a. Exception
    b. BaseException
    c. ExceptionBase
    d. Exceptions

11. A *try-except* block has
    a. At least one except
    b. Zero except
    c. Only one except
    d. None of these

12. What is the output of the following code?
```
def abc():
    try:return 'a'
    finally:return 'b'
print(abc())
```

  a. a
  b. b
  c. ab
  d. None of the above

13. What is the output of the following code?
```
Assert False, 'Hello'
```
  a. Hello
  b. True
  c. False
  d. None of the above

14. Which of the following is not a keyword in exception handling?
  a. accept
  b. except
  c. try
  d. else

15. What is the output of the following code?
```
lst = [1,2,3]
print(lst[3])
```

  a. Value error
  b. Index error
  c. Name error
  d. 3

16. What is the output of the following code?
```
a = '5'
b = '5'
print(a+b)
```

  a. Value error
  b. Type error
  c. Name error
  d. None of the above

17. What is the output of the following code?
```
Int('5.5')
```
  a. Value error
  b. Type error
  c. Name error
  d. None of the above

18. What is the output of the following code?
    ```
    def getMonth(m):
        assert m>1 or m<12, "Invalid"
        print(m)
    getMonth(6)
    ```

    a. Invalid
    b. AssertionError: Invalid
    c. 6
    d. None of the above

19. What is the output of the following code?
    ```
    x = Input("Enter a Number")
    print(x)
    ```

    a. ValueError
    b. NameError
    c. SyntaxError
    d. None of the above

20. What is the output of the following code?
    ```
    x = input("Enter a Number")
    print(X)
    ```

    a. ValueError
    b. NameError
    c. SyntaxError
    d. None of the above

DESCRIPTIVE QUESTIONS

1. What is an error? Which type of error is addressed at compile time?
2. What type of errors are addressed at runtime? Explain with an example.
3. What are the differences between *try-except* and raise expression statements?
4. List out five built-in errors and explain with examples when these errors occur.
5. Explain the exception hierarchy. Explain which among the below two statements should be used:
   (i) First use the BaseException and then ConcreteException.
   (ii) First use the ConcreteException and then the BaseException.

PROGRAMMING QUESTIONS

1. Write a function to concatenate two strings. The function takes two arguments. We do not know the type of the first argument in advance but the second argument is a String. Write a function that takes an unknown input and a String as input and tries to handle the error when we try to concatenate

this unknown input to the String using the try …except…else clause. The unknown input could be either an Integer or a String or a Float. If the concatenation fails, your function should return the value None (exactly without the quotes). If successful, your function should return the resulting String.

2. Write a function that divides two numbers. The function takes two arguments. The first argument is a Float but we are unsure about the second argument (there is a chance that the second argument could be a zero). Write a function that takes a Float and an unknown input and tries to handle the error when we try to divide the Float by the unknown input using the try …except…else clause. The unknown input could be either an Integer or a String or a Float. If the operation fails, the function should return the value None (exactly without the quotes). If successful, your function should return the result.

3. Write a function that modifies the content of a list. The function takes three arguments. The first argument is the list itself; the second argument is an index n and the third argument is a String. Set the nth (index) item of the list as the given String and return the modified list if successful. In case of a failure return the original list. Write a function that performs this task using the try …except…else statements.

4. Write a program for matrix addition. The program should notify if the order of the matrix is not the same.

ANSWER TO MULTIPLE CHOICE QUESTIONS

1.	A	2.	D	3.	C	4.	B	5.	B
6.	B	7.	D	8.	A	9.	A	10.	A
11.	A	12.	B	13.	D	14.	A	15.	B
16.	D	17.	A	18.	6	19.	B	20.	B

Index